RENEWALS 458-4574
DATE DUE

WITHDRAWN
UTSA Libraries

ANALYSING PROFESSIONAL GENRES

Pragmatics & Beyond New Series

Editor:
Andreas H. Jucker
(*Justus Liebig University, Giessen*)

Associate Editors:
Jacob L. Mey
(*Odense University*)

Herman Parret
(*Belgian National Science Foundation, Universities of Louvain and Antwerp*)

Jef Verschueren
(*Belgian National Science Foundation, University of Antwerp*)

Editorial Address:
Justus Liebig University Giessen, English Department
Otto-Behaghel-Strasse 10, D-35394 Giessen, Germany
e-mail: andreas.jucker@anglistik.uni-giessen.de

Editorial Board:
Shoshana Blum-Kulka (*Hebrew University of Jerusalem*)
Chris Butler (*University College of Ripon and York*)
Jean Caron (*Université de Poitiers*); Robyn Carston (*University College London*)
Bruce Fraser (*Boston University*); John Heritage (*University of California at Los Angeles*)
David Holdcroft (*University of Leeds*); Sachiko Ide (*Japan Women's University*)
Catherine Kerbrat-Orecchioni (*University of Lyon 2*)
Claudia de Lemos (*University of Campinas, Brasil*); Marina Sbisà (*University of Trieste*)
Emanuel Schegloff (*University of California at Los Angeles*)
Paul O. Takahara (*Kobe City University of Foreign Studies*)
Sandra Thompson (*University of California at Santa Barbara*)
Teun A. Van Dijk (*University of Amsterdam*); Richard Watts (*University of Bern*)

74
Anna Trosborg (ed.)
Analysing Professional Genres

ANALYSING PROFESSIONAL GENRES

Edited by

ANNA TROSBORG
The Aarhus School of Business

JOHN BENJAMINS PUBLISHING COMPANY
AMSTERDAM/PHILADELPHIA

The paper used in this publication meets the minimum requirements of
American National Standard for Information Sciences — Permanence of
Paper for Printed Library Materials, ANSI Z39.48-1984.

Library of Congress Cataloging-in-Publication Data

Analysing professional genres / edited by Anna Trosborg.
    p.   cm. -- (Pragmatics & beyond, ISSN 0922-842X ; new ser. 74)
    Includes bibliographical references and index.
    1. Business writing. 2. Technical writing. I. Trosborg, Anna, 1937- II. Series.
HF5718.3.A52   2000
808'.06665--dc21   99-058978
ISBN 90 272 5089 8 (Eur.) / 1 55619 921 X (US) (alk. paper)   CIP

© 2000 – John Benjamins B.V.
No part of this book may be reproduced in any form, by print, photoprint, microfilm, or any
other means, without written permission from the publisher.

John Benjamins Publishing Co. • P.O.Box 75577 • 1070 AN Amsterdam • The Netherlands
John Benjamins North America • P.O.Box 27519 • Philadelphia PA 19118-0519 • USA

Library
University of Texas
at San Antonio

Contents

Introduction

Anna Trosborg vii

Genre, Terminology, and Corpus Studies

Margaret Rogers
  Genre and Terminology 3

Intralingual, Interlingual and Intercultural Studies of Genres

Charles Bazerman
  Singular Utterances: Realizing Local Activities through
  Typified Forms in Typified Circumstances 25

Philip Shaw
  Towards Classifying the Arguments in Research Genres 41

Dacia F. Dressen and John M. Swales
  "Geological Setting/Cadre Géologique" in English and French
  Petrology Articles: Muted Indications of Explored Places. 57

Ines-A. Busch Lauer
  Titles of English and German Research Papers in Medicine
  and Linguistics Theses and Research Articles 77

Genres and the Media

Torben Vestergaard
    That's not News: Persusive and Expository Genres in the Press   97

Anna Trosborg
    The Inaugural Address   121

Genres in Conflict

Vijay Bhatia
    Genres in Conflict   147

Birgitte Norlyk
    Conflicts in Professional Discourse: Language, Law and Real Estate   163

Genres and New Technology (changing and emerging genres)

Greg Myers
    Powerpoints: Technology, Lectures, and Changing Genres   177

Lars Johnsen
    Rhetorical Clustering and Perceptual Cohesion in Technical (Online) Documentation   193

Broadening the Perspective

Susanne Göpferich
    Analysing LSP Genres (Text Types): From Perpetuation to Optimization in Text(-type) Linguistics   227

Index   249

Introduction

ANNA TROSBORG
The Aarhus School of Business

Full participation in disciplinary and professional cultures demands informed knowledge of written genres. Genres are the media through which scholars and scientists communicate with their peers. Genres are intimately linked to the discipline's methodology, packaging information in ways that conform to a discipline's norms, values and ideology. Understanding the genres of written communication in one's field is, therefore, essential to professional success.

At the same time, genre knowledge refers to an individual's repertoire of situationally appropriate responses to recurrent situations - from immediate encounters to distanced communication through the medium of print, and more recently, the electronic media. This volume also includes genres relying on the spoken medium, which has so far been overlooked in genre analysis.

The notion of genre originates from literary studies, where genres such as novels, short stories, poems, plays, etc. have been studied for centuries. In rhetorical studies, genre analysis has been carried out for almost twenty years. The focus has been on attempts to develop taxonomies or classificatory schemes or to set forth hierarchical models of constitutive elements of genre, which enable us to make generalisations about genre's form, substance and context (for reviews see, e.g. Miller 1984, Swales 1990, Yates and Orlikowski 1992).

Scholars in literary studies, rhetorical studies and in discourse analysis as well focus on the formal characteristics of texts rather than on the activities or practices in which genres are embedded. Their view rests on what Brandt (1990) called "strong text" (formalist) assumptions rather than a dialogical view of language-in-interaction (see Nystrand and Wiermelt 1991). However, a traditional rhetorical approach does not enable us to determine anything about the ways in which genre is embedded in the communicative activities of the members of the discipline or how the actor draws upon genre knowledge to perform effectively.

Genres and genre knowledge can be more sharply and richly defined to the extent that they are localised (in both time and place). A dialogical view is taken by sociolinguists and educational psychologists who share Baktin's (and Vygotsky's) view that social intercation is at the center of language and concept learning. Berkenkotter and Huckin (1995: 1-24) see genres as "inherently dynamic rhetorical structures that can be manipulated according to the conditions of use", and for this reason, genre knowledge is best conceptualized as "a form of situated cognition embedded in disciplinary activities". In their socio-cognitive theory of genre, they focus on five major aspects: *Dynamism, situatedness, form and content, duality of structure* and *community ownership*. These five aspects are very central to the framework of this volume.

The dynamic aspect has already been stressed. Genres are "dynamic rhetorical forms that are developed from actors' responses to recurrent situations" and "genres change over time in response to their users' sociocognitive needs." Even the scientific journal article, which has been perceived as a conservative, relatively static genre, especially on the formal level, has been found by Berkenkotter and Huckin to undergo significant changes when observed over a 45-year period. Experimental results are being increaingly foregrounded in titles, abstracts, introductions, and section headings, whereas methods and procedures sections are being increasingly relegated to secondary status. The main reason for this is to be sought in the information explosion, in which readers of scientific journals cannot keep up with the literature and are forced to skim journal articles the way many newspaper readers skim newspapers.

Genres are "a situated form of cognition" and as such our knowledge of them is derived from and embedded in our participation in the communicative activites of daily and professional life. Genre knowledge is "inextricably a product of the activity and situations in which it is produced (Brown, Collins and Duguid 1989: 33)". As the intellectual content of a field changes over time, so must the forms used to discuss it. This is why genre knowledge involves both form and content. Furthermore, as we draw on genre rules to engage in professional activities, we "constitute social structures and simultaneously reproduce these structures". The use of rhetorical genres is both constitutive of social structure (as it is instantiated through our observing a genre's rules-for-use or conventions) and generative as situated, artful practice. In using the genres customarily employed by other members of their discourse community, disciplinary actors help constitute the community and simultaneously reproduce it. For this reason, genres themselves reveal much of "a discourse community's norms, epistemology, ideology, and social ontology" (Berkenkotter and Huckin 1995: 4).

Studying genres, we observe the relationship between available patterns for communicative utterances and people's ability to alter or modify such patterns. Genres are sites of contention between stability and change. As products of dynamic societies, they are influenced by social structures, by changing social systems and not least by the rapidly growing technological development in modern society. As dynamic constructs, genres change with changes in society, and new genres emerge as a product of new technology. This creates a continual need for research in the changing nature of old genres and the emergence of new ones.

This volume extends our knowledge of professional genres including genres which have not previously been analysed. It stretches from genre and terminology to applied genre analysis. It is concerned with genres in different fields: economics (taxation), genetic engineering, academic writing, animal nutrition, geology, petrology, medicine, linguistics, journalism, politics, law, promotional literature (advertising, job offers), translation and business discourse, and it contains articles concerned with different media (spoken and written) as well as texts produced by multimedia.[1] The volume provides intralingual, interlingual, and intercultural studies. It compares closely related genres (theses and research articles, reports and leading articles, it provides intralingual as well as interlingual interdisciplinary studies (English/ French geology articles, English/ German research articles in medicine and linguistics). It takes up types of texts not previously investigated within genre analysis as diverse as written tax forms and the inaugural address. In addition to providing genre specific characteristics, a point is also made of highlighting the distinction between genre and text type (see Schäffner; Vestergaard). While German-speaking countries have maintained a distinction between *Testsorten* (genres) and *Texttyp* (text type), this distinction has often been blurred in other communities. Genres refer to whole texts often defined by their communicative purposes, whiles the notion of text type is concerned with properties of a text, in particular the way a text is built up by means of, for example, narrative, descriptive, expository and/or argumentative structural patterns.

The volume has a strong focus on socio-cognitive aspects. Genres are embedded in social action "as typified forms of typified circumstances" (see Bazerman, this volume for an extensive Anglo-American reference network). Genres are the intellectual scaffolds on which community-based knowledge is constructed (Berkenkotter and Huckin 1995: 24). To be fully effective in this role, genres must be flexible and dynamic, capable of modification according to the rhetorical exigencies of the situation. At the same time, they must be stable enough to capture those aspects of situations that tend to recur. Existing genres are adapted and new genres arise out of social conflicts, social changes, invention of new technology, etc. The tension between stability and change lies at the heart of genre use and genre knowledge and is dealt with in particular in the studies of genres in conflict (Bhatia; Norlyk) and in Myers' study of adaptation to new media. Finally, the perspective is further broadened with visions for future analysis. In the following, a brief outline of the individual contributions is presented.

The discipline known as terminology science is concerned with the specialist terms of LSP domains, the relations between them and the concepts which they designate. While texts have often played a part in terminology work, principally as a means of accessing expert knowledge, their role at the theoretical level has been much less clear, even deprecated, since text is language-in-use whereas terminologies seek to systematise and even normalise in some subject fields. *Margaret Rogers* argues that genre is a linking factor between traditional terminology science and the study of LSP texts. Using computer-based methods, she seeks to focus attention on aspects of text which cast light on the behaviour of terms not only as word forms, but also as potential markers of genre. The starting point for her claim that the behaviour of terms in a text is of a different nature from that of systematised terminologies (e.g. as found in dictionaries or term bases) is a study of compound formation in English and German genetic engineering articles. While genre, as a concept which has a classificatory role, is an important means of structuring collections of texts - or corpora - in order to facilitate the interpretation of terminological data, terms have also been viewed as a classifier of genre. The relationship between terminology and genre is often expressed using the metaphor of levels, whereby the same concept may be designated differently according to the level of abstraction of the text genre (e.g. learned article - professional article - manual - marketing brochure). However, a study of synonyms in an English corpus of automative engineering texts shows that different genres are less clearly characterised by the distribution of the synonyms themselves than by certain formal properties such as number of components in multiword compounds. The relations between genres are usually represented by another metaphor, that of hierarchy. Using the example of an LSP corpus of texts compiled for the purpose of building a terminology, and drawing on the computational metaphor of a "virtual corpus", Rogers argues that genre is a multifaceted concept consisting of attributes with various values which can be rearranged according to purpose. She concludes that the search for a fixed and exhaustive taxonomy of LSP genres is therefore elusive and even illusory, given the many different purposes which functionally-oriented genre classifications may serve.

Rhetorical approaches to genre bring together the typification of form and situation with the originality of each utterance. *Charles Bazerman* develops a preliminary understanding of rhetorically situated meaning within distinctive genres. The primary example of income tax forms (along with several secondary examples) reveals how ontologies of genre-appropriate objects to be represented within documents are formed and limited. The objects presented within the genre are frequently translated from other discursive systems or from practices that inscribe experiences; these translated objects must then appear in forms appropriate to the genre. The intertextual links to other documents and inscription practices are part of the means by which the meanings in a text are held accountable to representations outside the text. Once represented within the texts, these discursive objects may then be conceptually aggregated, related, and manipulated according to genre-appropriate operations.

The article by *Philip Shaw* examines the argumentative elements of the discussions sections of three theses and four research articles in Animal Nutrition. On the basis of surface linguistic features a classification is developed, dividing minimal arguments into seven types. The overall distribution of argument types was similar in both genres, but 'deductive' arguments occurred almost exclusively in the dissertations and evaluative expressions were much more frequent in the grounds of 'justify method' arguments in dissertations than in articles. More arguments in dissertations than in theses were chained into long sequences. These features reflect functional differences between the genres. Deductive and action-justifying arguments make frequent use of conjunctions like *since, therefore, if, then, thus*, etc, and are rarely hedged, while the argumentative vocabulary in the deductive and abductive arguments is dominated by expressions of implication like *suggest* and *indicate,* and hedges like *may* and *seem.* This supports a 'cognitive' rather than a 'social' interpretation of the phenomenon of hedging.

There has been considerable interest over the past 15 years in the scientific research article, its schematic structure and how it responds to discourse community expectations. The findings point to genre-characteristic features across disciplines. While these features have been widely investigated in numerous disciplines, relatively scant attention has been paid to the important discipline of geology. *Dacia F. Dressen and John M. Swales* set out to fill this gap. Morover, within the sub-discipline of petrology, they find a sub-section as yet unrecognized in the extensive literature on the research article. The 'Geological Setting', an introductory part-genre occurring before the onset of petrological analysis, is a multi-functional description which frames researchers' results within their geological context. Based on a corpus of 20 articles (10 in French and 10 in English), it is argued that this part-genre is not only a complex of the topography, geological history and characteristics of the research site, but also serves to establish the authors' credentials and authority as experts. This is accomplished not through agentive narratives marking the authors' presence on the site, but through more muted and indirect means. Similarities at the macro-level and general discourse structure are found in both the French and English texts, with important stylistic differences on a local level.

The article by *Ines-A Busch-Lauer* examines the structure, function and communicative effectiveness of titles in two subject areas - linguistics and medicine. The introduction surveys recent research on titles and outlines the functions of this genre in scientific discourse. The second part of the paper presents linguistic research results on 150 German and English titles (written by native speakers) randomly drawn from medical and linguistic specialist journals as well as conference proceedings. The titles were analysed with regard to their length, their syntactic structure and information content. A third corpus comprising L2-English titles written by German researchers was used to study similarities/differences in the approaches researchers apply in their mother tongue and L2. The study reveals interdisciplinary differences in the formulation and structure of titles: Titles in linguistics are rather short, vague, and abstract. Titles in medicine are long, precise and informative. Furthermore, the analysis indicates that German researchers use conventions of L1 to formulate titles in English, which often results in stylistically inappropriate titles.

In the ethos of the journalistic profession the distinction between "report" and "comment" is central and deep-rooted, and in spite of critical discourse analysts' attempts to show that it is ideological, *Torben Vestergaard* argues that the distinction has clear linguistic manifestations. For whereas news stories proper consist mainly of factual illocutions reporting what happened in the recent past and are careful to attribute other illocutions to sources other than the journalist, "comments", and leading articles in particular, freely contain non-factual illocutions such as evaluations, proposals, predictions, causal explanations, and interpretations, either unattributed or expressly attributed to the writer, the "editorial we". At the macro level, the leader is normally realized by expository, persuasive or hortatory texts, and as these text types can be regarded as points on a scale ranging from least to most urgent rather than discrete classes in a taxonomy, it would appear that there is a very clear and simple relationship between genre and text type in the case of leaders. This picture of a one-to-one relationship is disturbed, however, first by the relatively marginal phenomenon of "genre borrowing", where any text will be interpreted as a leader simply by virtue of occurring in the position in the newspaper where leaders are normally placed. There are, however, other non-news genres in a modern newspaper, and it is shown that texts which functional considerations would tempt us to subsume under the common generic label "background" can have so little in common that there is no justification for regarding them as representatives of the same type from a purely text-internal point of view. It is therefore argued that there is a need for upholding the conceptual distinction between "genre" and "text type", where the former is a functional term and the latter is based on textual micro and macro features.

The inaugural address of American presidents has been treated as a special genre involving special symbolic functions. *Anna Trosborg* is concerned with this genre and analyses in particular Cinton's 1993 address paying special attention to communicative functions and rhetorical strategies. On the basis of her findings, she points to specific characteristica of the address such as *confirming the presidential role*, *constructing the presidential identity* and speaking with *the official voice,* but at the same time she argues that the address shares many feautures with political speeches in general.

Two studies are concerned with genres in conflict. Genres have become dynamically complex as a result of several factors, which include, a tendency to mix 'private intentions' with socially recognised 'communicative purposes' (Bhatia 1993), a need to respond to complex, novel or changing socio-cultural contexts (Fairclough 1993), or complex demands of disciplinary cultures (Berkenkotter and Huckin 1995). While this often results in the mixing and embedding of genres, which fulfil somewhat non-conflicting and subservient communicative purposes, one may come across genres, which are called upon to serve conflicting demands of a particular profession. Based on the analysis of the *Joint declaration of the Government of the United Kingdom and the Government of the People's Republic of China on the question of Hong Kong,* *Vijay Bhatia* discusses issues arising from somewhat contradictory demands imposed on the same document, one to make declarations as certain of legal effect as possible and the other to be diplomatic and vague at the same time.

The article by *Birgitte Norlyk* focuses on two main issues within the field of professional discourse. The first issue concerns the problem of correct legal interpretation of a standard sales contract used in connection with the buying and selling of property in Denmark. In 1988, feeling the need to facilitate communication with customers, the Danish Board of Estate Agents presented a slightly modified version of the existing standard sales contract. The correct legal interpretation of this modified version subsequently caused a great deal of confusion within the Danish legal system. Thus, within the same year, the Danish Higher Courts - whose judgement normally constitutes a precedent for correct legal interpretation - gave different interpretations of the modified version of the standard sales contract used between 1988 and 1994. Today, the question of correct legal interpretation of the modified sales contract remains unanswered in the legal community. For practical purposes, however, the earlier version of the sales contract was reintroduced in 1994. The second issue concerns the questions of interpretation and professional communication within a broader framework. This part of the article highlights the problematic interaction between professional communication, professional discourse amd professional cultures.

New technological development influences existing genres, necessitating changes or it may give rise to new genres. The changes in one undergradute lecture on advertising regulation are used by *Greg Myers* to illustrate the relation between old genres and new technologies. The lecture has been revised to use the Powerpoint (TM) presentation package. But this was not just a change in format; it led to changes in the structure and reception of the lecture. His article explores the implications for genre theory, in treating the material form of texts, time, authorship, institutions, and embodied skills. The conclusion is that neither institutional pressures nor technological changes alone determine the development of genres; genres are better seen as resulting from the interactions of the various strategies of multiple actors.

In *Lars Johnsen's* article "Rhetorical Clustering and Perceptual Cohesion in Technical (Online) Documentaion" it is shown how professional communicators such as technical writers may achieve communicative goals not only through language but also through the interaction of the text and the visual design. The article focuses on the relationship between text design and textual organization in technical documentation. More specifically, it exemplifies how technical writers may create rhetorical clusters, Gestalts of text elements designed to work together as functional units, in technical (online) documentation and how the organization of these rhetorical clusters may be described on the basis of Campbell's principles of perceptual cohesion (Campbell 1995).

Genre knowledge may be utilized successfully in fields of application. Two well established areas of applied linguistics - translation studies and foreign language teaching both benefit from analysis within a genre perspective. *Christina Schaffner's* article falls within applied genre analysis. It deals with the application of genres for translation studies. The aim of any translation is to reproduce a target text that fulfils its specified purpose. After a short discussion of the concepts of text type and genre in textlinguistics research, it is illustrated how they have been applied in translation studies. For a large number of texts, translation as target text production means adapting the target text to the genre conventions of the target culture. Knowledge of cross-cultural similarities and/or differences regarding genres and genre conventions is therefore crucial to the translator. Such knowledge can be gained from a systematic analysis of parallel texts. The use of parallel texts and some other pedagogical implications for translator training are illustrated on the basis of job offers and news reports (English and German). The fact that not all the genres are equally influenced by conventional structures also puts limits to a translation-oriented parallel text analysis, which is discussed with reference to the genre of academic writing.

Finally, the perspective is broadened by the presentation of visions for future analyses. In the first part of her article "Analysing LSP Genres (Text Types): From Perpetuation to Optimisation in Text(-type) Linguistics", which refers extensively to German-speaking academic communities, *Susanne Göpferich* states the numerous LSP genres which have been analysed and compared intralingually as well as interlingually and interculturally so far need to be accompanied by an integration process in which the results of the individual analyses are joined together to form a mosaic that gives us an overview over larger text-type systems and insight into the relationships between LSP texts. She then proposes a procedure which may lead to such a mosaic.

In the second part of her article, she raises the question whether the characteristics of specific genres which have been found out on a purely descriptive basis actually contribute to the fulfilment of these genres' communicative functions in an ideal way. This assumption seems plausible, but, apart from a few exceptions, has not been verified. Therefore, Göpferich suggests that requirements to be met by the various genres should be defined (e.g. quick accessability of information, cognitive processability by specified groups of recipients, machine translatability, usability of instructive texts) and that empirical analyses should be carried out to find out whether the texts we are confronted with actually meet these requirements. Methods of such analyses and thus of investigating what ideal genres should look like are outlined in the last part of her article.

Notes

[1] For articles on new text types arising as a result of the tv medium, such as tv and dubbing, the reader is referred to the volume on *Text Typology and Translation*. (Trosborg (ed) 1995).

References

Berkenkotter, Carol and Huckin, Thomas N. 1995. *Genre Knowledge in Disciplinary Communication: Cognition/culture/power*. Hillsdale, New Jersey: Lawrence Erlbaum Associates, Publishers.

Bhatia, V.K. 1993. *Analysing Genre: Language Use in Professional Settings*. London: Longman, Applied Linguistics and Language Study Series.

Brown, J.S., Collins, A., and Duguid, P. 1989. "Situated cognition and the culture of learning". *Educational Researcher* 18: 32-42.

Campbell, K.S. 1995. *Coherence, Continuity, and Cohesion*. Hillsdale, New Jersey: Lawrence Erlbaum Associates, Inc.

Fairclough, N. 1993. *Discourse and Social Change*. London: Polity.

Miller, Carolyn R. 1984. "Genre as social action". *Quarterly Journal of Speech* 70: 151-167.

Nystrand, M. and Wiermelt, J. 1991. "When is a text explicit? Formalist and dialogical conceptions". *Text* 11: 25-41.

Swales, John M. 1990. *Genre Analysis. English in academic and research settings*. Cambridge, UK: Cambridge University Press.

Yates, J.A. and Orlikowski, W.J 1992. "Genres og organizational communication: A structurational approach". *Academy of management Review* 17: 299-326.

Genre, Terminology, and Corpus Studies

Genre and Terminology

MARGARET ROGERS
University of Surrey, U.K.

Introduction

This paper is concerned with the centrality of text to the study of terminology on both a practical and a theoretical level. Crucially, it is the notion of text that links terminology and genre. The paper is divided into three parts: terminology and documentation; genre and text corpora; and the relationship of terminology to genre.

Terminology and documentation

To start with, I would like to raise some issues which, in my view, give rise to some tensions between terminology science and what is implicit in the notion of "genre" as a system of text classification, and to try to resolve these issues by exploring the changing role of text in terminology theory and practice.

The conventional view of terminology science, particularly as presented in the Vienna school,[1] stresses the role of the concept as independent of the linguistic designation (Wüster 1974: 67), thereby distinguishing its approach from that of lexicology where form and content (i.e. meaning) are seen as a single lexical unit. The apparent independence of the concept on the epistemological or mental level and the term which is assigned to it (*sic*) on the linguistic level paves the way for some of the central tenets of terminology. These include the use of terminological principles in the standardisation of terminologies (Wüster 1974: 70-1, Picht and Draskau 1985: 178-92); the associated claim that the meaning of a term, i.e. the concept which it designates, is stable in any linguistic context (Felber 1984: 107-8); and the matching of characteristics of concepts in cross-linguistic work (Cole 1987). The postulated independence of the concept is crucial to the notion that language can somehow be controlled. Terminologists concede that control is inappropriate for general language but maintain that special languages are a different case, arguing that professional communication may be impeded by an uncontrolled growth and use of terms and the concepts they designate (Wüster 1974: 68, Felber 1984: 15). Terminologies, as codified special-language vocabularies organised according to a system of concepts, are the goal of terminology work. The terminology may then serve as a benchmark for use.

So far, so good. But what are the sources for such terminologies? And what is the relationship between a terminology and its sources? A terminology is an idealised representation of a special-language vocabulary in a given subject field, in essence consisting of lexemes (i.e. entry terms) and their associated concepts as realised in a system of definitions. In its normalising function in particular, it represents what might be called an authoritative and consensual view in which the essential characteristics of concepts and the most appropriate labels or "designations" have been agreed by a committee of experts. Documentation is widely acknowledged, and indeed recommended, as a source of linguistic and conceptual data (Picht and Draskau 1985: 166-7, Cole 1987). Yet as text, even carefully evaluated and selected documentation is language in use, consisting of word forms, not lexemes, and representing meaning by a complex web of textual relations which are often realised by lexical means. I am thinking in particular of the role of synonymy, hyponymy, paraphrase, repetition and partial repetition in establishing textual cohesion (Hoey 1991). Even a strict evaluation and selection of documentation cannot change the fact that a terminology - descriptive or prescriptive - is an attempt to represent the system of the specialist lexicon whereas text - evaluated or not evaluated - is language in use. And even a text written in accordance with terminological standards is still a text. Recent work by Gerzymisch-Arbogast (1996) on terms in context, for instance, has highlighted some systematic differences between the construction of meaning for terms in special-language texts and in codified sources such as specialist dictionaries or term banks. So while there are indeed qualitative differences between texts written for different purposes and with varying degrees of authority (e.g. journalistic *versus* academic), even authoritative texts often weave their meanings in ways which are systematically different from the way in which meaning is represented in codified sources such as terminologies and specialist dictionaries. Gerzymisch-Arbogast (1996) has shown, for instance, that for the subject field of economics, the use of terms in text by internationally-acknowledged experts varies with regard to the codified term-concept relation. Indeed, Riggs (1989) has pointed out that social scientists understand it to be part of their discourse to redefine or "resemanticise" key terms and concepts. But while we must allow for the possibility that subject fields such as those in the social sciences may be more susceptible to such variation than others, it seems unlikely that the creation of texts in the natural sciences is free from similar influences (cf. Ahmad 1997).

Seen in this way as language in use, text can no longer be regarded simply as a potentially confusing and somewhat tarnished realisation of what "ought" to be: it "is", and the LSP terms and their embedding in running text are simply a different type of "object" from the entry term and its associated data in an organised terminology. But in adopting this now not so new perspective, certain complexities in moving between use and system still need to be acknowledged.

General-purpose lexicography has always enjoyed a more comfortable relationship with its textual sources than terminology, since the semasiological - i.e. word-based - approach adopted in lexicographical work uses the linguistic context of words to distinguish their meaning, whereas terminology work emphasises the primacy of the concept, in keeping with its onomasiological orientation. However, the precise role which documentation plays in terminology work remains unclear in so far as the processes by which the terminologist is supposed to extract information on concepts and their relations from the text remain inexplicit for the large part and empirically under-researched.

Lexicographers were keen to embrace in a practical way the growing trend of data-driven lexically-oriented corpus-based studies which emerged during the 1960s (cf. Halliday 1966, Sinclair 1966 for early examples of such lexical studies), coinciding with and facilitated by technological developments in electronic data processing and storage. The use of corpora in general-purpose lexicography, particularly in English, is now well established following the publication of the Collins Cobuild English Language Dictionary (Sinclair 1987). Even in Language for General Purposes (LGP), where classifications may be viewed as more subjective than in Language for Special Purposes (LSP), genre has always been a consideration in the organisation of corpora. The texts in the Brown corpus from 1961, for example, were classified as informative or imaginative and then sub-classified into nine and six sub-genres respectively, including press, religion, skills and hobbies, popular lore, etc. on the one hand, and general fiction, mystery and detective fiction, science fiction, etc. on the other hand (Francis and Kucera 1982). In other words, such organisation was viewed as important if textual data were to be appropriately structured and interpreted. Genre has continued to be one of the important issues in corpus design and is not surprisingly a central issue for LGP lexicography (cf. for example, Summers 1993).

The use of corpora in terminology work has taken longer to establish itself and is only now being acknowledged in the mainstream literature (e.g. Meyer and Mackintosh 1996). One of the principal reasons for this time lag has to do with the view of LSP text mentioned earlier as something to be controlled and fixed through the normalising role of terminologies: the idea that a variety of textual sources should form the basis of terminologies appears to turn traditional thinking on its head. Add to this the role of the computer, implying an impoverished role for the terminologist/terminographer and a certain scepticism is not surprising. But variation in the use of terms can be viewed as a rich source of data, open, as any data, to interpretation. What facilitates such interpretation is the classification of the corpus according to a number of attributes, in which genre plays its part. The crucial question arising out of a re-oriented perspective in which text is regarded as *sui generis* rather than as an imperfect reflection of an ideal system, is: what is the relationship between the terminology in text and in a codified system? In other words, what can textual data reveal about the behaviour of terms and how should we interpret and represent this in terminologies, which are in turn the basis for the creation of new texts?

The study of LSP texts in the quantities made possible by computer storage and processing can reveal, for instance, linguistic patterns which are not evident through manual processing. An example concerning the compounding possibilities of terms from the perspective of synonymy and equivalence will illustrate the point. The study is based on a small bilingual corpus of scientific articles (c.51,000 words) in the field of genetic engineering (cf. Rogers 1997).

One of the most frequent critical observations made of sci-tech texts concerns the occurrence of synonyms, which leads to particular problems for translators when trying to establish appropriate equivalents. Normalising terminology work seeks to eliminate such multiple designations on communicative grounds and to establish one-to-one equivalents for preferred terms. The first step is a descriptive one, i.e. to identify cases of synonymy in each language. Based on an initial reading of the genetic engineering texts and consultation with experts, the following two possible synonym sets were noted in the texts studied (Levy 1994):

Table 1: Synonym sets in German and English (genetic engineering)

German	English
Erkrankung	disease
Krankheit	disorder
Morbus	illness
Syndrom	syndrome

The relative frequencies of these potential synonyms as lemmas were then calculated; the results suggested that *disease* and *Erkrankung* are the preferred terms, and therefore potential equivalents. However, it turns out that a complex pattern of interdependencies is at work involving categorial and combinatorial factors, namely, number (e.g. *disease* versus *diseases*) and head/modifier role in multiword nominal compounds (e.g. *acquired disease* versus *disease progression*). Using a Key Word in Context or KWIC analysis of the English subcorpus (33,872 words), it transpired that *disorders* rather than its singular counterpart is likely to occur as part of a multiword term (75% of the occurrences of the lemma DISORDER), whereas either *disease* or *diseases* is almost equally likely to occur as part of a multiword term (27% and 31% of the occurrences of the lemma DISEASE respectively). So, according to the data, *disorder* and *disorders* do not occur frequently as single-word terms and any multiword terms tend to be plural, but *disease* and *diseases* participate almost equally in the formation of multiword compounds of which there are proportionally fewer than for the lemma DISORDER. Single-word terms formed from the lemma DISEASE are more common than single-word terms formed from the lemma DISORDER.

A similar analysis of the German subcorpus (17,057 words) shows that the plural forms *Erkrankungen* and *Krankheiten* are also more likely to appear in compounds than their singular counterparts. In terms of equivalence relations we can note various patterns. So, for example, for multiword compounds in which *disease* is the modifier (e.g. *disease process*), the German equivalent term is likely to be a single-word compound (e.g. *Krankheitsprozeß*). For multiword compounds where *disease* is the head (e.g. *degenerative disease*), the German equivalent term is likely to be a multiword compound (e.g. *degenerative Erkrankung*).

So it seems to be the case that the notion of synonymy is more highly constrained if understood as relations between word forms rather than as relations between decontextualised lemmas, particularly if we consider the implications for cross-linguistic equivalence. The use of electronic corpora may then be said to contribute to a more richly-informed terminology, although the representation of this complexly-interrelated data in codified form on paper or electronically remains an outstanding problem. In other words, it is unclear how such data can be accommodated within the traditional dictionary or terminological entry structure or how it can in any straightforward manner be used as the basis for the elimination of synonymy.

Genre and electronic text corpora

The aspect of electronic text corpora which has attained an almost competitive edge is size. The question: "How large is your corpus?" is like a personal challenge. In general language a respectable size is now considered to be 100 million words plus, following the "publication" of the British National Corpus (cf. Clear 1993), whereas 30 years ago, 1 million was considered to be a significant achievement (as in the American Brown corpus, and its British parallel, the London/Oslo/Bergen or LOB corpus) (Francis and Kucera 1982, Hofland and Johansson 1982). The question of corpus size in relation to LSPs is, however, of a different nature for both practical and theoretical reasons.

It is generally acknowledged that it is more appropriate to speak of special languages in the plural rather than special language as a common variety (cf. Fluck 1985: 111). Each special language is, from a terminological perspective, related to a particular domain or subject field. But estimating the number of subject fields and establishing the divisions between them is a contentious and problematic task. What seems clear, however, is that the estimated number of subject fields is growing: estimated at 300 by Felber in the 1980s (Felber 1984), but retrospectively at 5,000 in 1955 and at 30,000 during the 1990s by the International Information Centre for Terminology (Infoterm). Each new terminology in a different subject field requires a new corpus to be built, often from limited resources. By contrast, in LGP lexicography the same corpus can be reused or expanded according to purpose. The issue of representativeness and what it means in LSP terminography as opposed to LGP lexicography - linked to the issue "What is a domain?" - remains unresolved, although there are indications that the ideal corpus size for terminology work is much smaller than that required for lexicography (Ahmad and Rogers 1994: 142-3).

In practical terminology work, delimiting the domain is a task which needs to be completed at an early stage (cf. for instance Picht and Draskau 1985: 164) in order to establish the boundaries of the terminology, both laterally with respect to other domains and vertically in relation to level of generality (cf. Meyer and Mackintosh 1996: 260, 269). This is usually done according to very practical criteria determined by the profile of the end-user group (or what we might call the "readership") and the purpose of the commissioned terminology, as we shall see. Another way of viewing the vertical dimension is from the perspective of genre as a means of characterising various levels of LSP communication.

If, for the moment, what we understand by genres is "conventional forms of texts associated with particular types of social occasion" (Hatim and Mason 1997: 218), examples in an LSP context would include learned articles, text books, reports, manuals, contracts, brochures, advertisements, and so on. Such a classification would also be relevant for certain terminological purposes, providing information on, for instance, vertical synonyms appropriate to different interlocutors according to level of expertise. Thus a simple LSP corpus design for written texts showing two levels of organisation could be as shown in Figure 1:

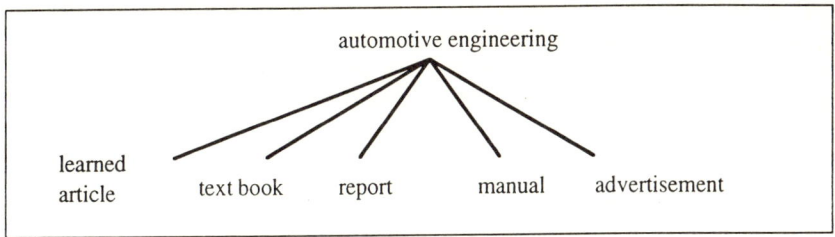

Figure 1: A simple LSP corpus design with two attributes (domain and genre)

But genres are themselves a complex of attributes[2] which may assume different values, including most notably author-reader relations (involving both competence and authority), function (e.g. expressive, informative, operative), channel (spoken/written) and conventional form. So advertisements, for instance, while always operative or persuasive in function and focussing on the reader, may not always assume the same form: a letter from a double glazing company may serve a similar promotional purpose to a brochure, lending support to Biber's view of genre as a classification determined by pragmatic criteria (such as situation or function) and text type as determined by text-internal criteria (Biber 1988: 170, Bex 1996: 6). A common solution when changing the values of key attributes of a genre is to talk of sub-genres. So the genre "manual" may be sub-divided according to variation in the attribute "readership" into "workshop manual" and "user manual", the former being aimed at workshop technicians, the latter at customers, who are laypeople. And so on.

The notions of genre and sub-genre, just as the notions domain and sub-domain, are slippery customers. There is, for instance, no general agreement on a classification or even an inventory of genres. As already indicated, the design of an LSP corpus is often related to the purpose of the terminological work in hand. So, if my purpose is to compile a trilingual translation-oriented terminology of sub-domains of automotive engineering for use in a particular company, then I would first consult the company to establish what kind of documentation is in use. The design of my corpus would then be constrained by the attributes domain/sub-domain, language, language variety, and whatever classification of texts could be established, possibly only implicit in working methods. As Gläser has commented (1993: 20), every text classification is "pragmatically founded", reflecting the fact that each one varies according to its "target" including in her view translators, LSP learners, or documentalists.

A corpus may therefore be viewed as a collection of texts which share certain attributes, but which may be arranged in different taxonomies according to user group and purpose. However, the texts of many corpora (e.g. Brown, LOB, Cobuild) are arranged in fixed taxonomies. If the texts which a particular user may wish to study are located in the same part of a taxonomy, they are easy to access. But another user may wish to study texts which are distributed through many parts of the taxonomy. This presents technical problems. Figures 2a and 2b show two possible taxonomies for the same set of English LSP texts in automotive engineering (AE):

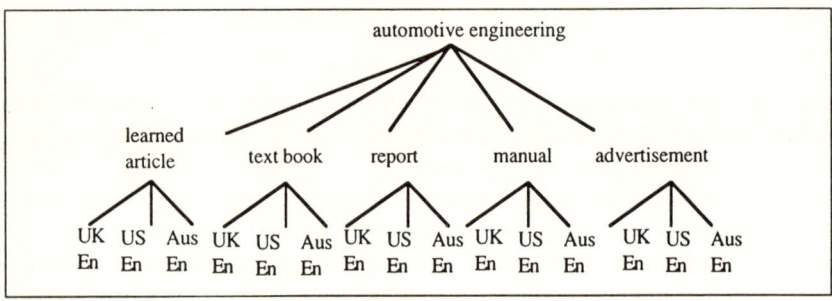

Figure 2a: Corpus arranged according to the hierarchy: domain, genre, language variety

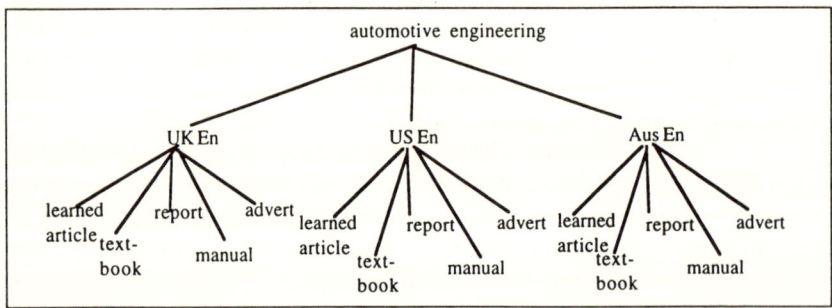

Figure 2b: Corpus arranged according to the hierarchy: domain, language variety, genre

The corpus design in Figure 2a would suit a genre-focussed study of terms in AE, whereas the corpus design in Figure 2b would serve a more language variety-focussed study. For instance, in order to compare the use of terminology in learned articles with that in advertisements, I only have to go to the second level of the tree in Figure 2a and can retrieve all the relevant texts together. I may then choose to process them in batch or individually depending on my purpose. To perform the same operation using the corpus in Figure 2b I would have to go to the third level and retrieve each set of texts separately. However, if my primary interest were genre variation within language varieties, then the organisation of the corpus in Figure 2b is more suitable.

The examples shown in Figures 2a and 2b feature only three attributes. If we add more attributes - such as date of publication, place of publication, gender of author, mother-tongue of author, original *versus* derived text, and so on - then the possibilities for reorganising the texts into many different taxonomies explode. And with it, the potential for lack of match with particular analytic purposes. So what we have is a very large number of potential taxonomies: little wonder that everyone's typology of texts or genres is different, given that their inventory of attributes also varies.

In fact, the nature of genre as a cluster of features or attributes has been used in the design of at least one general-language corpus, namely Longman's Contemporary Dictionary of English (Summers 1993), but the user can only build his or her sub-corpus by learning a query language. In order to provide a more user-friendly interface, avoiding both the use of a query language and the fixed nature of non-attribute based taxonomies, colleagues at the University of Surrey, Paul Holmes-Higgin and Khurshid Ahmad, have conceived and implemented a so-called "virtual taxonomy" or "virtual corpus" which allows a user to retrieve any sub-corpus (i.e. collection of texts according to selected attributes) from a given corpus (Holmes-Higgin, Ahmad and Abidi 1994, Holmes-Higgin 1995). Or, to put it another way, to allow any user of a corpus to be viewed as a "corpus designer" (Holmes-Higgin 1995: 18). While such a system ensures optimal re-use of the same material, e.g. for terminologies which have different specifications but which may be based in the same domain, it also embodies the fact that genre is a multifaceted concept for which an absolute definition eludes us. It also suggests that the very attempt at establishing a fixed taxonomy of genres may be illusory, given the variety of applications to which genre is relevant and from which its central notion of function is hard to separate.

The relationship between terminology and genre

One way in which LSP studies has in the past attempted to define something akin to genre is through the range of terminology used in sets of texts which are characterised by particular functions at different levels of professional communication. In so far as genre may be characterised in relation to textual function - i.e. according to external pragmatic criteria - we may draw a parallel here between the vertical layering of LSP and genre, as suggested above. Many schemas were developed from the 1960s to the 1980s in order to characterise different types of LSP communication, often according to a hierarchical notion of degrees of abstractness. The number of levels which have been proposed varies, usually between two and five, sometimes with further sub-divisions (Fluck 1985: 16-23 gives a good summary). Typically, the highest levels of abstraction are academic, e.g. the language of the theoretician and that of the experimentalist; the intermediate levels are related to the language of the practitioner, the technician, and the skilled shopfloor worker; and the lowest level of 'LSPness' (*Fachsprachlichkeit*) is characteristic of the language used to communicate with those outside the discourse community (e.g. the language of sales and marketing).

This is all well known. In reviewing this aspect of the LSP literature, I would, however, like to point to one important consideration: that all these classifications, regardless of their specific differences, relied in the first place on the specialist vocabulary or terminology of the domain to distinguish the different communicative layers (cf. Fluck 1985: 21-2). Indeed, such models are still current in the terminological literature. A very neat example of how terminology can be related to degree of abstraction is given in Arntz and Picht (1995: 19-20) who illustrate the five-layer model of Hoffmann using the domain of mechanical engineering:

Table 2: Hoffmann's five-layer model of LSP (vertical dimension) as illustrated by Arntz and Picht (1995: 19-20) from the domain of mechanical engineering (adapted from the German and condensed)

linguistic form	situation	interlocutors (bi-directional)	term
symbols for elements and relations	theoretical foundations	academic expert academic expert	$Sl=lBo-lWe$
symbols for elements and natural language for relations	experimental/ academic	technical expert technical expert technical assistant	$H\ 8/e\ 8$
natural language very high proportion of specialist terms	applied science and technology	technical expert production manager	*Spiel*
natural language high proportion of specialist terms	material production	production manager skilled worker master craftsman	*Luft*
natural language a few specialist terms	consumption	production team representative sales person consumer	*Spielraum*

But we should be careful how we interpret models such as that presented in Table 2: on the one hand, they indicate valuable generalisations, but on the other hand they may raise misplaced expectations of simplicity. One interpretation suggests that each specialist term has a less technical alternative - or vertical synonym - as shown in the last column of Table 2 and furthermore that particular terms only occur at certain levels of communication. In other words, the model raises neat expectations of a clear pattern of distribution such that the designations chosen for particular technical concepts vary between communicative levels.

The levels of communication shown in Table 2 can be related to genres as possible realisations of the types of communication envisaged at each level as illustrated in Table 3:

Table 3: Hoffmann's five-layer model of LSP (vertical dimension) as in Arntz and Picht (1995: 19) extended to show connection with genre

linguistic form	situation	interlocutors	genre
symbols for elements and relations	theoretical foundations	academic expert academic expert	*learned article (theoretical journal)*
symbols for elements and natural language for relations	experimental/ academic	technical expert technical expert technical assistant	*learned article (experimental journal)*
natural language very high proportion of specialist terms	applied science and technology	technical expert production manager	*article (professional journal)*
natural language high proportion of specialist terms	material production	production manager skilled worker master craftsman	*manual*
natural language a few specialist terms	consumption	production team representative sales person consumer	*marketing brochure*

So does the neat pattern which we may expect on the basis of such models accord with reality? These are, after all, in their expression as hierarchies and levels, only metaphors with which we attempt to understand reality. An analysis of a corpus of British English AE texts (c. 191,000 words) suggests that the real terminological picture is far from neat or straightforward. Table 4 shows the results of an analysis of nine terms which had been established as synonyms or variants (in the sense of denotational equivalents) in previous work, allowing us to investigate possible vertical variation in the use of designations for a particular concept. The distribution of each of the nine synonyms/variants is shown according to five genres:

Table 4: Distribution of nine synonyms/variants in a corpus of British English automotive engineering texts according to five genres in descending rank order of overall frequency

genre total words synonyms/variants	book 21,699 f	prof journal 48,774 f	popular science 14,456 f	newspaper 35,850 f	advert 35,985 f	190,950 N
catalyst	5	155	26	31	44	261
catalytic converter	25	61	32	37	56	211
autocatalyst		4			32	36
cat		9	3	1	17	30
catalytic convertor		2		2	1	5
automotive catalyst		2				2
catalysor				2		2
exhaust catalyst			1		1	2
exhaust catalytic converter					2	2
totals	30	233	62	73	153	551

What emerges from Table 4 is that the most frequent terms in books and journals, namely *catalyst* and *catalytic converter*, are the most frequent across all genres. What is more, there is evidence of what might be called permeation through the levels (continuing the metaphor) such that, for instance, less formal terms - in this case the abbreviation *cat* - appear in the upper levels. If we view these data from the perspective of genre, a picture emerges of a relatively broad distribution of terms across genres rather than an exclusive use of particular terms in particular genres. It is particularly notable that the less formal *cat* has the broadest spread after *catalyst* and *catalytic converter* as shown in Table 5:

Table 5: Range of distribution of synonyms/variants across genres

genre	catalyst	catalytic converter	catalytic convertor	cat	auto-catalyst	exhaust catalyst	exhaust catalytic converter	automotive catalyst	catalysor
book	x	x							
prof journal	x	x	(x)	x	(x)			(x)	
pop science	x	x		(x)		(x)			
newspaper	x	x	(x)	(x)					(x)
advert	x	x	(x)	x	x	(x)	(x)		

Note: (x): $f < 5$

For the chosen domain of AE, there does not therefore seem to be a clear association between the choice of terms and the communicative level for the nine synonyms identified. Rather than a system of levels, the analysis reveals a pattern which looks more like the representation in Figure 3:

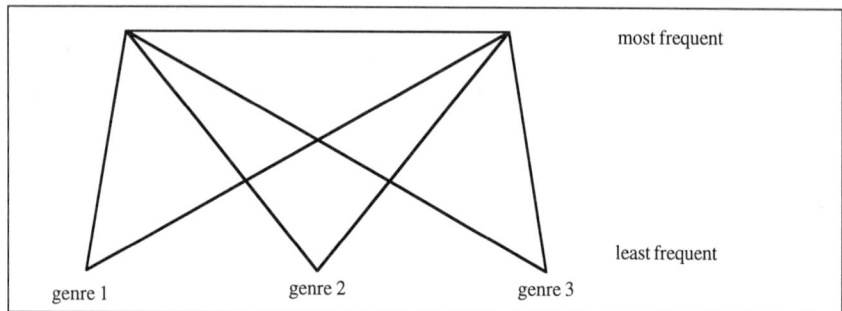

Figure 3: A model for the distribution of terms across genres according to frequency of occurrence

Another aspect of terminology which may distinguish genres is that of nominal compounding patterns, which are a typical characteristic of term evolution or formation. Here we might expect certain differences between more and less abstract texts in so far as compounds are often a formally very condensed way of representing complex concepts between experts. An analysis of compounds formed with *catalyst* revealed that in multiword compounds (N=268), in the majority of occurrences *catalyst* functions as the modifier e.g. *catalyst control, catalyst substrate* (f=173, i.e. 64.6%) rather than as the head e.g. *oxidation catalyst, pt catalyst* (f=95, i.e. 35.4%). A general tendency can be noted in the data[3] for *catalyst* to take on a modifier rather than a head function the less abstract the text. One explanation of this phenomenon is the nature of the compounds in which the key term is the modifier: the head of such compounds tends to be a sub-technical word (cf. Rogers 1997 for examples of the same phenomenon in genetic engineering). Examples from the AE corpus include: *catalyst technology, catalyst temperature, catalyst type, catalyst volume*, and so on. The tendency for key terms to function as modifiers rather than heads in multiword compounds can then simply be viewed as a reflection of the tendency to use more sub-technical terms at less abstract levels of communication.

Another formal aspect of multiword compounds is the number of components which they contain. In the present corpus, examples were found of compounds with up to seven components as in Table 6:[4]

Table 6: *Examples of multiword compounds containing "converter" and "catalyst"*

no. of components	converter	catalyst
2	converter geometry	catalyst technology
3	hc-co converter	catalyst ceramic substrate
4	single bed catalytic converter	single-bed oxidation catalyst
5	single-bed oxidation catalytic converter	automotive emission control catalyst technology
6	co and hc catalytic converter system	closed loop three way catalyst system
7	closed-loop-controlled three way catalytic converter	

It seems reasonable to assume, in the spirit of Zipf (cf. Zipf 1935/1965), that longer terms will occur less frequently than shorter ones. We might also assume that longer terms, which are more difficult for readers to process, occur less frequently in genres which are at lower levels of abstraction.

Table 7 shows that there is indeed a tendency for the length of compound to decrease with the loss of expertise in the readership. We can see that the incidence of lengthy compounds tends to decrease as the genre changes; the frequency of shorter compounds is also higher than that of longer compounds in the corpus as a whole:

Table 7: *Distribution of multiword compounds containing "converter" and "catalyst" according to genre*

no. of components in compound	Book f	Prof journal f	Popular Science f	Newspaper f	Advert f	Totals f
2	5 16%	95 53%	7 26%	16 41%	10 42%	133 44%
3	13 42%	66 37%	12 44%	16 41%	11 46%	118 39%
4	7 23%	13 7%	8 30%	5 13%	2 8%	35 12%
5	1 3%	3 2%	0 0%	1 3%	1 4%	6 2%
6	3 10%	3 2%	0 0%	1 3%	0 0%	7 2%
7	2 6%	0 0%	0 0%	0 0%	0 0%	2 1%
totals	31 100%	180 100%	27 100%	39 100%	24 100%	301 100%

To sum up, the relationship between terminology and genre reflects, as we might expect, communicative needs, but not necessarily as a straightforward vertical ordering of synonyms according to degree of abstractness. The data in the present paper suggest, for instance, that a formal characteristic of terms, namely the number of components in nominal compounds, is a clearer marker of genre than the choice of a particular synonym or variant. How far this result may be domain-dependent or domain-type dependent (e.g. applied technology) remains to be seen.

Conclusion

In this paper, I have attempted to show some ways in which the use of terms in text may be analysed using electronic LSP text corpora as a source and how genre plays an important role in structuring and interpreting terminological data.

An important consideration in the analysis of terminological data was shown to be the different nature of terms as used in text (word forms) and as represented in codified collections of terms (lemmas). This was illustrated through an analysis of compound formation in English and German genetic engineering terms.

Some ways in which terminology has been viewed as a classifier of genre have also been considered. The resulting picture is, not surprisingly, a complex one in which the metaphor of levels gives way to less familiar representations with regard to the distribution of synonyms across genres, and where the distribution of terms according to some formal characteristics is more consistent with the traditional metaphor.

Continuing with metaphors, the notion of a hierarchy in the form of a fixed taxonomy to represent relations between genres has also been reviewed in the light of the functional definition of genre and of the concept of a virtual hierarchy as proposed in the computerised management of corpora. It was concluded that the pursuit of a universal fixed taxonomy of genre was misguided in view of the many configurations made possible by the regrouping of attributes in accordance with a variety of analytic purposes (what is the taxonomy for?).

Terminology and text, text and genre, genre and terminology: following the links of this chain presents us with interesting new perspectives which are both informative for and informed by the development of computer-based techniques and concepts.

Notes

1. Cf. Wüster 1974, Felber 1984, Wüster 1985, Picht and Draskau 1985.
2. "However arbitrary may appear to be the boundary between one genre and another, what distinguishes them, and what determines how genres are traditionally defined, is usually the set or cluster of structural and STYLISTIC properties that have come to be associated with them, which have come to be DOMINANT in the FORMALIST sense; also certain TONES or attitudes, subject-matter, WORLD-VIEWS and audiences." (Wales 1989:206, author's own emphases).
3. Although there is a tendency for the proportion of modifier roles to be higher for the less abstract texts, e.g. 32/44 or 72.7% for adverts, as opposed to 20/32 or 62.5% for newspapers and 13/22 or 65.5% for popular science, journals also showed a higher proportion of modifier rather than head roles (108/165 or 65.5%). This is perhaps influenced by the fact that the journals in this corpus are of a professional rather than academic kind, exhibiting a lower degree of abstractness. The number of occurrences of multiword compounds with *catalyst* in books is low ($f=5$), but in all cases *catalyst* is the head of the compound.
4. Hyphenated elements have been treated as separate components since there seems no good reason to treat, for instance, *hc-co converter* or *hc/co catalyst* differently from *hc co catalyst*, which has been classified as a three-component compound.

References

Ahmad, K. 1997. "Deconstructing Knowledge and Scientific Writing: Acquisition and Representation of Knowledge." Plenary lecture. 11th European Symposium for Special Purposes, Copenhagen Business School, 18-22 August 1997.

Ahmad, K. and Rogers, M. 1994. "Text-based concept-oriented terminology - Eine Neuorientierung?" In A. Grinsted and B. Nistrup Madsen (eds), *Festskrift til Gert Engel i anledning af hans 70 års fødselsdag*. Frederiksberg: Samfundslitteratur, 134-47.

Arntz, R. and Picht, H. 1995. *Einführung in die Terminologiearbeit*. Hildesheim: Georg Olms. 3rd edition.

Bex, T. 1996. *Variety in Written English*. London and New York: Routledge.

Biber, D. 1988. *Variation across speech and writing*. Cambridge: CUP.

Clear, J. 1993. "The British National Corpus". In G.P. Landow and P. Delaney (eds), *The Digital Word: Text-Based Computing in the Humanities*. Cambridge, MA/London, UK: MIT Press, 163-87.

Cole, W. 1987. "Terminology: Principles and Methods". *Computers and Translation* 2: 77-87.

Felber, H. 1984. *Terminology Manual*. Paris: UNESCO.

Fluck, H. 1985. *Fachsprachen*. Tübingen: Francke Verlag. 3rd edition.

Francis, W.N. and Kucera, H. 1982. *Frequency Analysis of English Usage: Lexicon and Grammar*. Boston: Houghton Mifflin.

Gerzymisch-Arbogast, H. 1996. *Termini im Kontext. Verfahren zur Erschließung und Übersetzung der textspezifischen Bedeutung von fachlichen Ausdrücken.* (Forum für Fachsprachen-Forschung. Band 31). Tübingen: Gunter Narr Verlag.

Gläser, R. 1993. "A Multi-level Model for a Typology of LSP Genres". *Fachsprache*, 15(1-2): 18-26.

Halliday, M.A.K. 1966. "Lexis as a linguistic level". In C.E. Bazell, J.C. Catford, M.A.K. Halliday and R.H. Robins (eds), *In Memory of J.R. Firth.* London: Longman, 148-62.

Hatim, B. and Mason, I. 1997. *The Translator as Communicator.* London: Routledge.

Hoey, M. 1991. *Patterns of Lexis in Text.* Oxford: OUP.

Hofland, K. and Johansson, S. 1982. *Word Frequencies in British and American English.* Bergen: Norwegian Computing Centre for the Humanities.

Holmes-Higgin, P. 1995. *Text Knowledge: The Quirk Experiments.* Guildford, University of Surrey. Unpublished PhD Dissertation.

Holmes-Higgin, P., Ahmad, K. and Abidi, S.S.R. 1994. "A Description of Texts in a Corpus: 'Virtual' and 'Real' Corpora." In W. Martin, W. Meijs, M. Moerland, E. ten Pas, P. van Sterkenburg and P. Vossen (eds), *EURALEX '94: Proceedings of the 6th EURALEX International Congress on Lexicography.* Amsterdam: The Netherlands, 390-402.

Levy, I. 1994. *"Das Prinzip der Gentherapie" and "Molekularbiologie in der Neurologie". A Translation with Commentary.* Guildford: University of Surrey. Unpublished MA Dissertation.

Picht, H. and Draskau, J. 1985. *Terminology: An Introduction.* Guildford: The University of Surrey.

Meyer, I. and Mackintosh, K. 1996. "The Corpus from a Terminographer's Viewpoint". *International Journal of Corpus Linguistics* 1(2): 257-85.

Riggs, F. 1989. "Terminology and Lexicography: Their Complementarity". *International Journal of Lexicography* 2(2): 89-110.

Rogers, M.A. 1997. "Synonymy and Equivalence in Special-language Texts. A Case Study in German and English Texts on Genetic Engineering". In A. Trosborg (ed), *Text Typology and Translation.* Amsterdam: John Benjamins.

Sinclair, J. M. 1966. "Beginning the Study of Lexis". In C.E. Bazell, J.C. Catford, M.A.K. Halliday and R.H. Robins (eds), *In Memory of J.R. Firth.* London: Longman, 410-30.

Sinclair, J. M. (ed) 1987. *Looking Up: An Account of the Cobuild project in lexical computing.* London and Glasgow: Collins.

Summers, D. 1993. "Longman Lancaster English Language Corpus: criteria and design". *International Journal of Lexicography* 6(3): 181-208.

Wales , K. 1989. *A Dictionary of Stylistics.* London: Longman.

Wüster, E. 1974. "Die allgemeine Terminologielehre - ein Grenzgebiet zwischen Sprachwissenschaft, Logik, Ontologie, Informatik und den Sachwisenschaften". *Linguistics* 119: 61-106.

Wüster, E. 1985. *Einführung in die allgemeine Terminologielehre und terminologische Lexikographie*. Copenhagen: The LSP Centre, UNESCO ALSED LSP Network, The Copenhagen School of Economics.

Zipf, G. K. 1935/1965. *The Psycho-Biology of Language: An Introduction to Dynamic Philology*. Cambridge, MA: MIT Press.

Intralingual, Interlingual and Intercultural Studies of Genres

Singular Utterances:
Realizing Local Activities through Typified Forms in Typified Circumstances

CHARLES BAZERMAN
English Department, University of California, Santa Barbara

Genre and utterance

Every time we write, we create a new utterance for a new circumstance. That's why writing is so hard: each time we write we have to think of new, appropriate, effective words in an extended turn as part of an interaction that is not immediately visible to us, an interaction we have to imagine. On the other hand, we write in identifiable realms of discourse, mobilizing recognizable forms to locate our activity, perceive possibilities, shape intentions, and make our utterances intelligible to our readers. That's what makes writing not totally impossible or unimaginable. Theory and research on genre help us identify the invented social spaces that mediate communication.

When we think of originality in typified locales, it is perhaps the idiosyncratic examples that come to mind — such as Gould and Lewontin's "The Spandrels of San Marcos and the Panglossian Paradigm," originally delivered at a Symposium of the Royal Society, and published in the *Proceedings of the Royal Society* (Gould and Lewontin 1979). This ostensibly scientific paper, offering a critique of dominant evolutionary thought, mixes quotations from Voltaire and architectural drawings of medieval cathedrals with a review of evolutionary literature going back to Darwin. Such examples often blur, overlay, or otherwise twist genres into new shape. Or perhaps we think of singular utterances as the singularly successful text that appears thoroughly ordinary, but says something that turns out to be extremely important, such as Watson and Crick's "A Structure for Deoxyribose Nucleic Acid." (Watson and Crick 1953)

We ought to remember, however, that the most ordinary and undistinguished article in any journal has something to say, is the product of extensive work, and attempts to intervene in some novel way in an ongoing discussion—is, in short, an utterance.

Text, discourse, rhetoric and composition

We often have a hard time bringing together our notions of typified, genred regularity of writing with our appreciation of the novelty and specificity of each new utterance. In the study of the rhetoric and discourse of science (the LSP area I am most familiar with) we can see two opposite approaches towards individuality and regularity, but each equally serve to keep novelty and genre far apart.

One cluster of work aims to demonstrate the individuality of utterance, arguing that scientific and technical writing is a skilled, local activity, a matter of art and therefore human construction. Motives for this work range from the appreciation of individuals (as in Locke 1992), to revealing the nature of the art (as in the essays in Selzer 1993), to calling into question the epistemic authority of an objective science (as in Collins 1985).

Another cluster has been to find the regularized forms and processes of language, pragmatics, and text organization. Here the motive has been primarily the study of language with an eye towards education and text improvement, so as to teach students the ways of language they will need to read and write. This work includes Halliday and Martin's (1993) examination of nominalization, Swales' (1990) work on article introductions, and Myers' work on irony and politeness (1989, 1990). Investigations into these patterns has been pursued through large corpuses and through singular examples, but even the singular examples are studied as to how they either reveal general patterns or have entered into the historical production of regularities. Sometimes historians of science also consider how individuals have worked within communicative regularities that we no longer share, to remind us how different the pursuit of knowledge was in other times and places (as in Biagioli 1993).

Between these we might place work with a rhetorical impulse — that is, a concern with the strategic use of the regularized processes and resources of communication. This approach mixes concerns for the particular and the general to provide practical advice for framing utterances and evaluating the utterances of others. Prelli (1989) and others have pursued the rhetoric of science through principles of classical rhetoric which assume similarity of processes, techniques and resources across all situations. However, classical rhetoric's subsumption of special purposes into general advice washes out the particularities of science, technology, or any specialized endeavor, in ways that both misguide and perhaps alienate practitioners who are motivated to participate in a special practice. Practitioners are only too aware of the particularity that distinguishes their endeavor, and that provides perhaps their very reason for preferring their mode of professionalism. And for first and second language learners, the particularity of one set of practices can provide the motivation and direction for more effective and advanced learning. Thus classical rhetoric is of limited value from an LSP perspective. (See Bazerman and Russell (1994) on the persistent historical resistance of rhetoric to considering specialized languages and situations.)

A different rhetorical approach that does attend to some of the particularity of science and technology has been Bruno Latour's political vision of scientific persuasion, presented in *Science in Action* (Latour 1987). He sees the regularities of scientific texts (such as citation patterns, use of numbers, and appeals to the laboratory) and of scientific exchanges (such as raising the stakes in debates so that investigators with lesser resources have to drop out) as resources in trials of strength. He shows how discourse creates alliances and solidifies the strengths of particular communicative networks in ways that make those strengths invincible, pervasive, and invisible.

Latour's account serves as both a critical and a productive rhetoric — that is it helps you to see through the tactics of others and to produce your own tactics. Latour's rhetorical savvy accounts for much of his popularity in science studies, for people find in his work a real feel for how argumentative struggles go in any competitive discourse arena. Several limitations of his approach, however, are consequential for LSP — namely, in attributing too much influence of the heroic individual in reshaping social and communicative relationships, in not attending adequately to the slowly evolving structures of social interchange, in seeing all communicative relations as agonistic power struggles, and in providing an awkward and insufficient way for considering how human discourse is responsive to the resources and pressures of the nonhuman. These limitations direct attention away from genre and other regularities in the form, social organization, and interaction that would help students orient to and learn how to participate in special purpose communications. Further they direct attention away from the representational potential of special languages; that is, by casting relations between text and non-textual actants as totally driven by the need to create power alliances, actant network theory provides little to help us understand meaning or how readers and writers find meanings in texts.

What has been called the North American approach to genre (Freedman and Medway 1994, Russell 1997) provides a way to consider the development of historically evolved specialized social forms in relation to individual actions deploying these forms in concrete historical moments. This tradition of research and theory recognizes that genres are always remade by each individual's novel action and that the discursive spaces within recognized genres create opportunity spaces for individual utterance within ordered social relations and activities. Thus this approach provides means to consider how the specialized languages of disciplines and professions offer the means to make novel contributions to historically unique conversations.

This theory of genre began by linking the rhetorical tradition of genre studies to Schutz's phenomenology of everyday life through the concept of social typifications (Schutz 1967, Schutz and Luckmann 1973, Miller 1984, Bazerman 1988). The linkage between socially evolved, socially recognized forms and individual sense-making and self-expression allows a link between formal approaches to language and the long expressivist tradition of composition studies. We now can recognize more clearly that one learns to express oneself in particular circumstances in particular social fields through recognizable social forms. Identity becomes realized on specific social stages, even if there is a personal backstage that looks out to the several venues of public production.

The North American tradition of genre studies has developed means for considering the emergence and transformations of textual forms; the social roles and interactions mediated through these social forms; the ideological, epistemological, and communicative assumptions realized in these forms; the reading and writing processes associated with particular genres (Bazerman 1988, 1993, 1994, 1997a, 1997b); the persuasive resources of various genres and mixed genres (Journet 1993) the kinds of knowedge expert users of genres deploy in participating in genres (Berkenkotter and Huckin 1995, Freedman 1993, Freedman, Adam and Smart 1994, Prior 1998, Blakeslee 1997); the relations among genres within professions (Devitt 1991, McCarthy 1991, van Nostrand 1994, 1997) the relation between genred textual practices and other non-textual aspects of professional practice (Yates 1989, Schryer 1993) and many other related social phenomena. However, this approach has developed less fully and systematically issues relating to the specific meanings represented within the textual space of genres. That is, while individual analyses of texts have discussed the content of the analyzed texts in relation to the genre, they have not developed genre-relevant concepts for considering how genre shapes representational content. Two major exceptions are the study of specialized persuasive topics embodied within genres (initiated with Miller and Selzer

1985); and the role of taxonomy in establishing meaning categories, particularly with respect to clinical psychology (McCarthy and Geiring 1994; Berkenkotter and Ravotas 1997; Ravotas and Berkenkotter forthcoming).

In the remainder of this paper, I suggest several genre-influenced meaning creating processes that allow us at the micro-level to utter original, situation relevant but still genre appropriate representations, namely: populating space with objects, translation from other systems and discourses, intertextuality, accountability, and operations. To explain what I mean by these concepts I will use as my primary example a creative, individualized utterance within a highly typified, regulated, and coercive discursive field, where there would seem to be little room for self-articulation—the income tax report. This extreme example can show us in stark form how individuality of utterance might play out in other genres, from which I will draw secondary examples.

The concepts I present here have some family resemblance to what Halliday (1982) calls the ideational aspect of text, what Searle (1969) calls the representational act within speech acts, and what classical rhetoric characterizes as the logos of a speech. However, the concepts here specifically provide ways of considering how genre-shaped discursive spaces constrain and provide opportunity for the representation of particular kinds of genre-relevant meanings.

Regulation and uniqueness

Tax forms are a remarkable kind of self-confession where one reveals intimate details of life to strangers, details that one would not share with friends. After all this confession, one writes a check and mails it off to one's government. Now that is a powerful form of writing. Of course the document does not achieve its self-punishing confessional power entirely by itself as an abstract literary text; it is surrounded by laws, records, accounting systems, criminal justice systems and other resources and contexts that make one accountable for reporting specific information, calculating according to procedures, correlating representations with other orders of representation, and otherwise being mindful of what one says on the tax page. Constrained and directive as the form may be and constraining and compulsive as the surrounding contexts may be, people spend much time thinking through what they will write, and then spend much additional money to hire consultants, accountants, and lawyers to help prepare their self-representations. Great emotions and anxieties may surround the writing of these confessions and awaiting the response of readers in the tax office.

People feel tax forms are where they are most reduced to a faceless number, but tax forms are also where one's most full representation occurs, one's must full self-articulation — work, affluence, home, family and dependents, charity giving, extra income producing activity and income producing wealth, extraordinary health expenses, travel, and a variety of other activities that might influence one's tax liability. A major strategy for avoiding difficulties is to make oneself in fact non-noticeable, to obscure one's exceptionality or particularity, keeping all one's deductions within standard (but unspoken) guidelines.

This self report is dialogically constructed with the makers of the forms who identify the major categories and terms of self-representation. Much effort and expense goes into the construction of the forms. The dialogic construction is even more complex because one would assume that the evolution of the form is at least in part in relation to the clever stratagems of previous respondents as well as users' complaints and confusions. Moreover, the experience of reading and evaluating prior responses enters into continuing revision of the form and thus the self-representation it produces.

Generic ontologies and unique objects

Consider the standard United States Individual Income Tax Return, form 1040, for the 1996 tax year (see figures 1 and 2, Appendix pp. 39, 40). First let us consider the category of objects which will populate this discursive universe. Each genre facilitates representation of particular universes of objects, and the particularity of any utterance has in part to do with the particular objects of the appropriate kind one chooses to place in that represented universe. We might call these the ontology of the genre, and the specific ontology of each utterance.

One compulsory object for all filers, reprinted on the top of forms, as in figure 1, is the government. Another object (required by fill-in spaces near the top of the form) is the individual filer (or couple filing jointly) — the tax payer. One can try to keep oneself out of the universe by not filing, or one can file under an alias or otherwise try to disrupt the connection between one's embodied person and the person textually represented in the universe. There are categories of children and adults who need not file, usually based on income, and it is sometimes possible to remain invisible to the Internal Revenue Service (IRS). However, there are intertextual and other means of making you accountable for providing an accurate name. Further, addresses representing residences are part of the ontology because the IRS wants to attach the named filer with a body resident at an address, to correlate this document with other documents related to that person as well as to make that embodied person directly accountable for the symbolic representations made in his or her name.

Moreover, the IRS would like to make the ontology as stable as possible — the same people from year to year, identifiable as the same people, statistically likely to be living at the same address — so they provide pre-printed stick-on labels to correlate the person from one year with the next (see figure 1 near the top). Social Security numbers also enter the ontology in the attempt to make the people represented in each year continuous with people in previous and future years. Social Security numbers also establish identities within another system of record-keeping and tax-taking.

There are some legal choices of self-representation within this part of the ontology (that is, thoroughly accountable variations that will not be policed by a variety of coercive means). If one has multiple addresses, one can make a choice, subject to various criteria, of a primary residence. One can also choose to be single, married, or married filing separately. Also there is some flexibility in identifying additional people, characterized as dependents.

Most of the objects requested on these forms are numbers representing money, in turn representing categories of income and expenses. Many of these categories are specifically identified and are directly tied to other documents filed with the government (see, for example, line 7 of figure 1 and line 52 of figure 2). But some categories are more open, based on what you believe you can report without calling undue attention to yourself or for which you believe you can make yourself accountable if called on to do so; for example, gains or losses on rental real estate, farm income, or moving expenses. For some of these elective categories of income and expenses, you may be immediately asked to provide some greater account on additional forms. On these further forms you have some rhetorical choice in how you characterize various items and under what category you will place them. You may even add persuasive descriptions to argue for the deductibility of the item; for example, "travel to conferences and professional meetings" or "home office for preparation of manuscripts and lectures." These elliptical arguments through description can be translated into more extended arguments carried out in tax courts and the appellate court system.

In other domains there are other ontologies, as I have recently become aware of in my study of the discourses surrounding the emergence of incandescent lighting and central power into late nineteenth century America (Bazerman forthcoming). For example, patents have ontologies of inventions, often first introduced as illustrations and then described as components of the illustrated object, inventors, geographic locales, and dates.

In newspaper interviews with Edison, of course, Edison is necessarily part of the ontology, as is usually Menlo Park, but both person and place are frequently treated as wonderful or magic objects; further both are typically associated with other wonderful and magical objects of inventions, experiments, or just ordinary equipment transformed, which fill the geographic space of Menlo Park and come in contact with Edison's personal space. One might place this genre of feature interview stories within a larger genre of feature stories involving exotic and wonderful individuals residing in exotic places, but having some connection to ordinary life. One of the same newspapers that wrote Edison interview stories was sponsoring expeditions to the Arctic and to Africa, going to great expense to plausibly populate stories with wonder-filled exotica and heroic individuals. Further, when such an article appears in a phrenological journal, head size and shape also are a necessary part of the ontology.

Translation

Where do these objects that populate texts come from? Where do the numbers you fill in on the tax forms come from? Well, you could make them up, just making sure the linguistic form is plausibly appropriate. Names should consist of two words and optionally a middle initial, should resemble names within the diverse mix of the American population, and should not obviously appear to be an insulting phrase, the name of a basketball team, or anything else that might call attention to its inappropriacy. Income figures should appear in U.S. dollars expressed in decimal integer form, and again should not take on an extremity of value that would call attention to itself.

So entries must appear in the right form. If you conceive of them in some other form, you must recast them in ways that allow them to be received and recorded within the symbolic universe this document is supposed to produce. If, for example, you wish to insult the tax collector through the address, you must code your insult to look like an address. Income that appears in descriptive words (for example "a large quantity of gold") needs to be quantitized, translated into dollars, and then transcribed in decimal form. Charity giving in kind — such as used clothing given to the Salvation Army — is a typical place that exposes the nature of the translation process, as there is some leeway in the evaluation and who does it.

This is similar to the process in courts where certain objects from the lives of victims and accused are admitted into court, but only that which is admissible by the rules of evidence and translated into the appropriate form to be considered as evidence and entered into the persuasive and deliberative discourse enacted in the court. Similarly Latour and Woolgar (1979) talk about the process of inscription by which mice by a series of translations are transformed into data in a form appropriate to the scientific argument to be made.

The material in Edison's notebooks and the other inventions being cooked up in the lab need to be translated into the form appropriate to a patent. The patentable object need not come from a particular working successful technology, but only from an idea, so it is often easier to trace Edison's patents back to his notebooks than an actual working object or experiment in the lab.

Intertextuality

Most name and number entered on tax forms are supposed to come from particular other places, and you are held accountable by a variety of means to transcribe these accurately from the appropriate other places. For example, the name you inscribe should be the one on your social security, driver's license, last year's tax returns, and other legal documents.

Often where representations of objects in one system come from is from another text. The stubs on your paychecks or the records in your employer's account books are added up to produce an amount called annual wages. Tips do not have an intertextual trail; as a result frequent tensions arise between the IRS and waiters, street buskers, cab drivers and others who earn gratuities which are not easily and accountably aggregated. Travel expenses, similarly, may only reside in a variety of crumpled receipts which may or may not have been saved, perhaps supplemented by a personal log book. These intertextual resources for preparing your taxes are of varying rigorousness and compulsivity.

Writing an academic paper affords greater degrees of freedom in choosing which intertextual resources can be drawn on and displayed. Some of the idiosyncrasy of Gould and Lewontin's paper is just in their generically inappropriate reliance on such intertextual resources as Voltaire's Candide and analyses of Gothic cathedrals. But even without pushing the boundaries of citation practices of a genre, academic writers have great leeway in tying their work to various parts of the appropriate literature and drawing on various resources made available in the literature. These choices often serve to index the author's intellectual position among the alternatives in the field.

Accountability

In different genres and the surrounding discoursal practices we may be held accountable for the various representations we make within the generic utterance. So the Internal Revenue Service may make us specifically accountable for the sources of the numbers we enter. That is, we may be called into their offices in order to provide an account of where we got our numbers from. The kind of account we need to come prepared to give is an intertextual one, with all the relevant records and receipts. Financial and perhaps criminal penalties will result if we fail to come up with a good accounting and intertextual support.

The IRS to ensure accountability collects major parts of the intertextual record in advance, through employers' reports of earnings, statements of miscellaneous earnings, bank and stock broker reports of earnings, and similar filings provided to both the government and the taxpayer. The IRS then correlates those income reports with the personal reports we file on our tax returns. In the central computer our tax returns are made intertextually accountable, and we are reminded if there is any lapse in our set of reported linkages — that is, if we neglect to report some income that someone else has reported giving to us.

The reports sent to the government have already done the work of translating, so that we have no options in how we will represent the intertextual information — rather we just transcribe the provided numbers in designated spaces and must even include copies of some of the intertextual mediating documents, such as the W-2 statement of earnings form. A major tax issue is, of course, what escapes this intertextual web of reporting and how the government can make us accountable for that. This area of irregularly reported transactions is known as the 'grey economy'.

In another example of intertextual accounting, from the Edison papers, patents are accountable not for being workable or effective or profitable, but only for being ideas had by certain people on certain dates, as attested to in the patent. So records of inventive thought, in the form of notebooks, are the appropriate form of intertextual account one can provide in appeals, hearings, and court cases. As a result the pages of the working notebooks are regularly dated, signed, and notarized. Further they are afterwards catalogued, annotated, and correlated with intermediate summary notebooks and with granted patents.

Other kinds of genres may be made accountable to other kinds of realities, but only through mechanisms that textualize or inscribe the non-textual materials so as to translate them into the discourse. Thus it is important for the success of Edison's project not only that he have successful demonstrations illuminating Menlo Park, but that these events were reported on in the newspapers and technical journals, establishing wide-spread and enduring meanings for the financial, corporate, scientific, and public worlds.

The mechanisms of inscription are not arbitrary or without meaning; they are the site at which the inscriptions can be made accountable. Genres that carry their force by appeal to the emotions or experience are held immediately accountable to the self-accounts of the readers — "This moved me; that other left me cold and rung false to human emotion." Statements that hold their force by their claim to be representative of the public will or the union membership are likely to be embarrassed by any misrepresentation or by the changing moods and occasions of group articulation.

Within scientific experimental and observational reports, method provides an account of the way in which inscriptions were produced and thus makes the link between the experiment or the observed event and the inscribed account. Not only must one tell a persuasive story of the method (that is, an account that meets current beliefs and expectations of how to turn events into data) but one is also accountable for having carried out these procedures as described, competently and without falsification, manipulation, or other deviance. After the fact investigations into researchers' ethical conduct may pursue accountability by examining notebooks and searching for corroborating documents, by which experiments and observations were inscribed.

Accountability as well is a resource. One can add strength to one's utterance by increasing the kinds of accountability one opens oneself up to, by displaying mechanisms by which one is holding one's text accountable to various textual or non-textual existences, and by drawing on the strength or dynamic of particular areas considered of consequence by the readers. Thus a preacher in making a sermon accountable to the responsiveness of the congregation by including various eliciting techniques that require the co-participation of the audience can draw on that strength for even greater emotional force on that audience; early demonstrations of emotional response encourage further emotional response.

Concepts and operations

Once a text is populated with various appropriate objects appropriately translated into its represented world and fully accountable to carry the full force of the other realities represented, the text can then do things with these objects, can transform them further through operations upon the symbols. It can aggregate and divide them; it can show that they form a process or provide evidence for a more general claim; it can calculate consequences. Within each genre, there are certain appropriate operations the text can carry out, and certain domains of abstractions that are appropriate to invoke in carrying them out.

To use the stark example of the tax form, entered numbers are added and subtracted, thereby turning wages, dividends, rental income, alimony and so on, into 'total income' (see line 22 figure 1). The total income is then transformed into 'adjusted gross income' through further subtractions (see line 31, fig. 1). 'Adjusted gross income' is next transformed through 'computation of taxes,' and further recombined with 'credits,' 'other taxes,' and 'payments' to determine 'refund' or 'obligation.' Similarly, people can be transformed into 'dependents' and then into 'exemptions,' which in turn have dollar values in the calculations (see figure 1, lines 6a-6d; figure 2 line 36. The various operations lead to conclusions, and the concepts used along the way serve both to define the operations and to identify the higher level entities that result from the operations.

Similarly in patents, the specifics of the object get transformed into generalizable statements of ownership of ideas, known as the 'claims.' These higher level entities of 'claims' become one's intellectual property — but not until the examiner at the patent office carries out operations of evaluation, approval, issuing the patent, and assigning a patent number, again transforming the claim into another kind of entity designated by the patent number.

As another example of the creation of concepts through textual operations, in the *Bulletin of the Edison Electric Light Company*, which served as an early marketing document, anecdotes of fires and suffocation caused by gas lighting were regularly published to invoke fear of the competition; these were set alongside accounts of the safety of electric lighting and approvals of electric lighting by insurance underwriters. These stories were aggregated to build a concept of safety that would be uniquely attributable to electric lighting.

The exploratory and incomplete concepts I have sketched out provide tools to think about how representations are fostered, constrained, and shaped by various genres as well as to consider how those representations are made parts of purposeful individual utterances. Just because we write in genres, and try to speak to the expectations of others, creating recognizable mediating spaces for communication, this does not at all mean we give up our opportunity to create novelty and speak to our moments. Rather, it is those mediating spaces that give us the very means to utter ourselves into new modes and moments of being, to represent ourselves and the worlds we live in.

References

Bazerman, C. 1988. *Shaping Written Knowledge: The Genre and Activity of the Experimental Article in Science*. Madison: University of Wisconsin Press.

Bazerman, C. 1993. "Money Talks: Adam Smith's rhetorical project". In W. Henderson, T. Dudley-Evans and R. Backhouse (eds), *Economics and Language*. London: Routledge, 173-99.

Bazerman, C. 1994. *Constructing Experience*. Carbondale, IL: Southern Illinois University Press.

Bazerman, C. 1997a. "The life of genre, the life in the classroom". In W. Bishop and H. Ostrom (eds), *Genres of Writing*. Portsmouth NH: Boynton Cook.

Bazerman, C. 1997b. "Discursively structured activities". *Mind Culture and Activity* 4(4).

Bazerman, C. forthcoming. *The Languages of Edison's Light*. Cambridge: MIT Press.

Bazerman, C. and Russell, D. R. 1994. "The rhetorical tradition and specialized discourses". In C. Bazerman and D. R. Russell (eds), *Landmark Essays in Writing across the Curriculum*. Davis CA: Hermagoras Press, xxi-xxxxviiii.

Berkenkotter, C. and Huckin, T. 1995. *Genre knowledge in disciplinary communication: Cognition/culture/power*. Hillsdale, NJ: Erlbaum.

Berkenkotter, C. and Ravotas, D. 1997. "Genre as tool in the transmission of practice over time and across professional boundaries". *Mind Culture and Activity* 4(4).

Biagioli, M. 1993. *Galileo, Courtier: the Practice of Science in the Culture of Absolutism*. Chicago: University of Chicago.

Blakeslee, A. M. 1997. "Activity, context, interaction, and authority: Learning to write scientific papers in situ". *Journal of Business and Technical Communication*, 11: 125-169.

Collins, H. 1985. *Changing Order: Replication and Induction in Scientific Practice*. Beverly Hills: Sage.

Devitt, A. 1991. "Intertextuality in tax accounting". In C. Bazerman and J. Paradis (eds), *Textual Dynamics of the Professions*. Madison: University of Wisconsin Press, 336-357.

Freedman, A. and Medway, P. 1994. "Locating genre studies: Antecedents and prospects". In A. Freedman and P. Medway (eds), *Genre and the New Rhetoric*. London: Taylor & Francis, 79-101.

Freedman, A. 1993. "Show and tell? The role of explicit teaching in the learning of new genres". *Research in the Teaching of English* 27: 222-251.

Freedman, A., Adam, C., and Smart, G. 1994. "Wearing suits to class: Simulating genres and genres as simulations". *Written Communication* 11: 193-226.

Gould, S. J. and Lewontin, R. C. 1979. "The Spandrels of San Marco and the Panglossian paradigm: A critique of the adaptationist program". In J. Smith and R. Holliday (eds), "The Evolution of Adaptation by Natural Selection: A Discussion". *Proceedings of the Royal Society of London* B 205: 581-98.

Halliday, M. and Martin, J. 1993. *Writing Science*. Pittsburgh, University of Pittsburgh Press.

Halliday, M. 1982. *Language as a Social Semiotic*. London: Edward Arnold, 1982.

Journet, D. 1993. "Interdisciplinary discourse and 'boundary rhetoric': The Case of S.E. Jelliffe". *Written Communication* 10: 510-541.

Latour, B. 1987. *Science in Action: How to Follow Scientists and Engineers through Society*. Cambridge, MA: Harvard University Press.

Latour, B. and Woolgar, S. 1979. *Laboratory Life: The Social Construction of Scientific Facts*. Beverly Hills: Sage.

Locke, D. 1992. *Science as Writing*. New Haven: Yale University Press.

McCarthy, L. P., and Geiring, J. P. 1994. "Revising psychiatry's charter document, DSM-IV". *Written Communication* 11(2): 147-192.

McCarthy, L. P. 1991. "A psychiatrist using DSM-III: The influence of a charter document in psychiatry". In C. Bazerman and J. Paradis (eds), *Textual Dynamics of the Professions*. Madison: University of Wisconsin Press, 358-378.

Miller, C. R. 1984. "Genre as social action". *Quarterly Journal of Speech*, 70: 151-67.

Miller, C.R. and Selzer, J. 1985. "Special topics of argument in engineering reports". In L. Odell and D. Goswami (eds), *Writing in Nonacademic Settings*. New York: Guilford Press.

Myers, G. 1989. "The pragmatics of politeness in scientific texts". *Applied Linguistics* 10(1): 1-35.

Myers, G. 1990. "The rhetoric of irony in academic writing". *Written Communication* 7(4): 419-455.

Prelli, L. J. 1989. *A Rhetoric of Science: Inventing Scientific Discourse*. Columbia, S.C.: University of South Carolina Press.

Prior, P. 1998. *Writing/Disciplinarity: A Sociohistoric Account of Literate Activity in the Academy*. Mahwah, NJ: Erlbaum.

Ravotas, D. and Berkenkotter, C. forthcoming. "Voices in the text: Varieties of reported speech in psychotherapists' initial assessments". *Text*.

Russell, D. R. 1997. "Writing and genre in higher education and workplaces: A review of studies that use cultural-historical activity theory". *Mind Culture and Activity* 4(4).

Schryer, C. 1993. "Records as genre". *Written Communication* 10: 200-234.

Schutz, A. 1967. *The Problem of Social Reality*. The Hague: Martinus Nijhoff.

Schutz, A. and Luckmann, T. 1973. *The Structures of the Life-World*. Evanston, IL: Northwestern University Press.

Searle, J. 1969. *Speech Acts*. Cambridge: Cambridge University Press.

Selzer, J. (ed). 1993. *Understanding Scientific Prose*. Madison: University of Wisconsin Press.

Swales, J. 1990. *Genre Analysis: English in Academic and Research Settings*. Cambridge: Cambridge University Press.

Watson, J. D. and Crick, F.H.C. 1953. "A structure for deoxyribose nucleic acid". *Nature*: 737-738.

Van Nostrand, A. D. 1994. "A genre map of R&D knowledge production for the US Department of Defense". In A. Freedman and P. Medway (eds), *Genre and the New Rhetoric*. London: Taylor & Francis, 133-145.

Van Nostrand, A. D. 1997. *Fundable Knowledge: The Marketing of Defense Technology*. Mahwah, NJ: Erlbaum.

Yates, J. 1989. *Control through Communication: The Rise of System in American Management*. Baltimore: Johns Hopkins University Press.

Figure 1

Figure 2

Towards Classifying the Arguments in Research Genres

PHILIP SHAW
The Aarhus School of Business

Introduction

In this paper I want to approach genre from the point of view of text type, and specifically to start from what has been said about argumentation. I use the term *text type* as van Dijk does, to refer to a limited number of large-scale discourse types such as narrative, expository, directive, descriptive, and argumentative. van Dijk (1992) says these types are 'characterised in semantic and schematic terms' and refers to argumentative schemata as parallel to narrative schemata, seeing them as composed of ordered sequences of elements like premises and conclusions. Genres typically include stretches of text definable, for example, as narrative, and other sections definable as argumentation or exposition. The boundary between these two last categories is not always clear. It may not even be necessary: Martin's categories (1989) of 'Analytical Exposition' and 'Hortatory Exposition' seem to cover the same ground as van Dijk's exposition and argumentation, and thus to dispense with the notion of argumentation. I shall therefore not attempt to distinguish persuasive and expository argumentation at this stage, though my results do point to possible formal differences (see below).

The object of study here is the argumentative elements (in van Dijk's sense) of two academic genres — dissertations and research articles. van Dijk describes argumentation as a textual structure characterised by propositions some of which are in the semantic relation of 'support' to others. Consequently a minimal argument, from a textual point of view, consists of two propositions, one supporting the other. *Swans are white* is in this sense not argumentative; *All swans are white, because I've never seen a black one* is an argument.

Large sections of research-reporting genres like the dissertation or the research article are argumentative in the sense that they draw conclusions from data and seek to persuade the reader that some proposition is true. Among propositions of other types there are minimal arguments (pairs of propositions related by 'support'), and they often play key roles.

Several accounts of argumentation include typologies of arguments (deductive, inductive, abductive, with subtypes of induction like *modus ponens*) and of argumentative schema, ways in which people argue in ordinary life. One useful contribution from argument theory to describing research genres might be a typology to which formal differences could be linked. Linguistically the minimal arguments have a variety of forms, and the aim of this paper is to examine the possibility of a (pilot) classification of these forms. It investigates whether argument typologies can differentiate the arguments in scientific writing in a way which illuminates the genres of the scientific article and the dissertation, and possibly helps to distinguish them. Argument in scientific writing is particularly worth investigating because as argument in LSP one would expect it to be characterised by typification, codification, and restriction, and thus to include a limited number of argument types, making analysis a realistic task.

If a typology of argumentative schemes could be made applicable it could be revealing in three different ways. First, different types of argument might reliably use different linguistic forms. This would be useful information in itself, and conscious knowledge of it would enable novice writers to avoid miscuing readers.

Second, one might expect to find different types of argument in different sections of the conventional introduction-method-results-discussion report and thus to be able to give a more explanatory account of the differences between the sections. Brett (1994) and Thompson (1993) observe, for example, that early cycles in the Results/ Discussion section frequently justify features of method or analysis, perhaps on the basis of types of results. It would be interesting to know at what point in Discussion sections minimal arguments of different types occur.

Third, one might expect to find that different types of argument occur across genres even in the same field. Research genres will have different patterns of argument from textbooks (Myers 1992) but it is also possible that dissertations exhibit different types of argument from 'professional' research articles either because the genres have different purposes or because their writers have different levels of skill.

Previous studies

The most frequently cited analyses of Discussion sections in research reports do not make use of argument-structure in the description. Table 1 shows Hopkins' and Dudley-Evans' (1988) influential move-sequence analysis of the discussion section (based on applied biology dissertations), which fits the data I shall examine very well but appears to me to miss some structure.

Table 1. Hopkins' and Dudley-Evans' Discussion section moves

Background information
Statement of result
(Un)expected outcome
Reference to previous research (comparison)
Explanation of unsatisfactory/surprising result
Deduction
Hypothesis
Reference to previous research (support)
Recommendation
Justification

Except for Statement of Result, the moves are optional and tend to occur in cycles rather than linearly, so that the Discussion section will be a series of result statements each followed by further material of varying kinds. Hopkins and Dudley-Evans do not discuss relations among members of a cycle, although their terms Deduction and Hypothesis imply an argumentative relation; nor do they discuss relations among cycles. Argument structure provides a way of discussing these relations; it fits in at a level between the individual moves and the cycles, showing functional relations among the moves, and also establishing inter-cycle relations, in that the outcome of one cycle may be part of the input to a higher argument.

It is a natural assumption that there is argument structure in results/discussion sections. There have been a number of previous studies employing argument theory in analysing such sections. Most make use of Toulmin's terminology (Toulmin, Rieke and Janik 1984). Thompson (1993) sets out a basic model for the structure of results sections in biochemistry which is a useful guide to the issues involved.

Table 2. Thompson's model

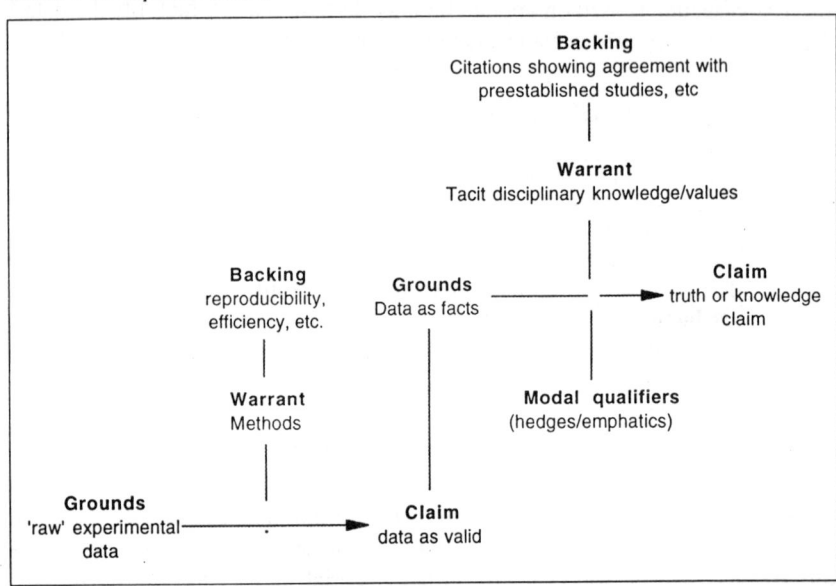

As Table 2 shows, Thompson's model involves two levels of claim. The first is a claim that the data to be used are valid or reliable, and this is normally warranted by the description of the method. There is usually no argument in the sense defined above at this stage, though Myers (1990) shows that in a controversy it becomes necessary to argue for the validity of one's data. The stage which I am interested in is the second one, where the validated data become the grounds or premises for a claim or conclusion. Thompson uses Toulmin's terms in such a way that the implicit knowledge which connects the grounds and the claim is the warrant, backed where appropriate by parallel or similar results from other work.

An interesting study which investigates the development of generic competence is Chou Hare and Fitzsimmons (1991). They used the Toulmin model to compare attempts to write a discussion section by undergraduate, master's level, and doctoral students of education.

They expected the more sophisticated writers to organise their arguments into quite long chains, so that the conclusion of one argument was a premise of the next. However in fact they did not find this phenomenon frequent in any of their subjects' texts. This perhaps reflects the tendency observed by Bazerman (1994) to conceal explicit argumentation in research reports.

They also conceived of the ideal discussion section in their field as having claims based on data, with warrants and backing made clear. Demands for precision and explicitness in technical writing might well lead one to expect such 'complete' arguments. In fact however, the well-structured discourse community (Swales 1990, Bizzell 1993 etc.) in which the arguments are presented probably provides readers with shared knowledge, and so the conditions for incomplete argument. Grice's principles make it very hard not to be cooperative and seek the missing premise, and where a reasonable amount of knowledge is shared it is possible to find it (Vestergaard 1994). Thompson shows that warrants in scientific writing are usually tacit, though it is also quite usual to give the backing for a warrant in the form of a reported parallel finding. It is thus not surprising that Chou Hare and Fitzsimmons found that, although doctoral students produced discussion sections closest to their ideal, with claims based on data and with interaction between them, even here warrants for claims were usually lacking.

An aim of this paper is to take Chou Hare and Fitzsimmon's findings one stage further, and assess the differences, if any, between the actual dissertations of PhD students and published articles in terms of explicitness and chaining.

Design of this investigation

The procedure was in two parts. First I collected a small set of texts, became familiar with their content, and identified minimal arguments as defined above (that is, pairs of propositions, one supporting the other), and their relationships with each other (chaining, etc.). Then I developed a preliminary classification, applied it, and asked a number of research questions about the output of the classification.

The collection consisted of research reports in the field of Animal Nutrition, a branch of applied biology or physiology. There were three extracts from dissertations which were set out in a pattern which distributed Discussion into several sections. I took my extracts from the arguments which appeared in the discussion of results and of methodological problems within the experiments rather than on the 'General Discussion', so that my data corresponded more closely to earlier than to later phases of the discussions in Hopkins and Dudley-Evans' model.

The other four texts I looked at were the Results and Discussions sections of four research articles from the *Canadian Journal of Animal Nutrition*.

The two groups were chosen to get an idea of the sort of differences there might be between PhD discussions and those in articles, either because they are different genres or because (at least in the case of the present texts) the writers of dissertations are less experienced.

Example of arguments

Table 3 illustrates two successive cycles from one of the published articles, the second of which includes two minimal arguments. The terminology used for the moves is that of Hopkins and Dudley-Evans, but I have drawn attention to a

Table 3. A sample cycle including two minimal arguments

Section Title	Serum Urea Nitrogen
Cycle 1	
Statement of results = data as facts	At 30, 37,60, and 67 d. of age, SUN [serum urea nitrogen] concentrations did not differ between EW [early weaned] and LW [late weaned] lambs (Table 5).
Result label	These SUN values
Reference to previous research (comparison)	are within levels reported in suckling (Bassett 1974:Lane and Albrecht 1991) and weaned lambs (Lane et al 1988; Cole et al 1988; Lane and Albrecht 1991).
Cycle 2	
ARGUMENT 1	
Input/results = data as facts	Serum urea nitrogen concentrations were greater ($P <0.05$) at 60 and 67 d. of age than at 30 and 37 d. of age in EW lambs. Similarly, in LW lambs, prefeeding values (hour = 0; Fig. 1 were higher ($P <0.05$) at 60 and 67 d. of age than at 30 and 37 d. of age. Postfeeding SUN concentrations continued to differ ($p>0.05$) up through hour 4.5, after which values were similar among days of age.
Result label (ground)	The differences in SUN concentrations between 30 and 60 d. of age in EW and LW lambs
Claim 1	may reflect an increase in amount of CP consumed
reference to previous research (warrant+backing)	since SUN levels are positively correlated to protein intake (Preston et al 1965).
ARGUMENT 2	
Higher ground label (=claim1 below)	This
Higher claim (Claim 2)	would suggest that SUN concentrations were not affected by the dietary protein source (milk vs pelleted feed) in this study.

particular noun-phrase type which often occurs at the transition from the statement of result to the next part of the cycle - the data label. As noted by Francis (1994) labels of this kind can serve to evaluate the material they summarise or to highlight the significant aspect. In some cases they provide a link with a different tradition of argument analysis by providing an argumentative orientation (Ducrot 1980) to the ground. That is, choice of label (*these large differences, this variation*) predisposes readers with the necessary background to a particular claim.

The two arguments are in series with the second building on the first ('chained' in Chou Hare and Fitzsimmon's terms). The first is 'complete' in the sense that it includes ground, claim, and backing for the tacit warrant, while the second is more typical in only having ground and claims, and thus being enthymematic (like most real arguments (Vestergaard 1994)). The difficulty which most of us have in reading the text and filling in the missing steps reflects its orientation to a particular discourse community.

Developing a surface-feature classification

It would be possible to use many criteria for the classification of surface arguments, because they differ from one another in many different ways. In order to start thinking about this, I had to develop a typology which might be heuristically revealing and provide the basis for a theoretically better-founded future investigation. I therefore aimed for a small set of applicable and distinct categories, however incoherent the criteria.

Argument theorists normally find that the surface forms of 'naturally-occurring' arguments are misleading and one has to have recourse to 'reconstructed' forms which represent the real arguments irregularly represented in the surface forms. However, these reconstructed arguments are underdetermined by the surface and there are several possible logical forms for the abstract argument. Although only one of these is likely to be discoursally sound (Vestergaard 1994), any typology of the arguments in an LSP genre has to be based on the surface arguments; the 'reconstructed' forms involve too much inference for application.

It might be possible to distinguish between the surface logic of the text and the linguistic forms it is presented in, but there is a considerable risk that the surface logic is merely an interpretation of the linguistic form. If one classifies the sentence *A major aspect that may account for the different responses in the two studies could be the protein levels in each trial* as a claim about causation, is one responding to the surface logic or simply to the word *account for* ? To avoid this difficulty, the typology proposed here, although making use of 'logical' labels from Walton and van Eemeren and Grootendorst, is based on linguistic characteristics of the text sections in question. The labels and categories are designed to be interesting from an argument-theory point of view but do not represent a claim that the arguments examined can be reconstructed in the informal-logic category hinted at.

A classification of the surface features of minimal arguments

I first of all identified all the pairs of propositions that seemed to me to be in an argumentative relation. This was inevitably quite subjective and the resulting list may not be exhaustive. I then intuitively identified important types, defined them – objectively — in terms of surface features, and counted the members of each type.

One type — JUSTIFY ACTION — is defined by having in its claim an action of the writer/researcher. These can be related to van Rees's (1992) *Problem-solving* — at least they consist of a situation and the action taken because of it.

(1)   ...in LW lambs, day of age x hour of sampling interactions (P<0.05) were observed for serum insulin...and SUN; therefore, *values were examined* within hour of sampling. (article)

(2)   Unfortunately the haemocrits were not determined during this trial and an assumed value *was adopted* to convert BF rates to plasma flow rates. (dissertation)

In another — STATUS — the claim consists of an expression summarising results, the copula, and an expression which evaluates/characterises the result in some way. A justifying expression was always present, though not criterial.

(3)   These yields.....are not *unexpected* because peak lactation often occurs near 30 d postpartum (Wohlt et al 1984). (article)

(4)   Considering the individual variation between the sheep....(e.g .....Wolff et al 1972.....), any conclusion based on the mean of only four sheep must be *treated with caution*. (dissertation)

In the remaining arguments the surface claims were statements about the object of research. In a large group which I classified as ABDUCTIVE-CAUSE the claim consists of a result summary statement of some kind, an expression of cause, and another expression referring to an attested phenomenon or process which explains or causes the results. These seemed similar to Walton's (1989) inductive arguments of causation. The type is called abductive here because it seems to call on a generalisation (the warrant, unexpressed in 5, induced from the literature in 6) to connect two specific statements.

(5) A major aspect that *may account for* [expression of cause] *the different responses in the two studies* [expression summarising results] could be the protein levels in each trial [explanatory phenomenon]. (article)

(6) *This* [expression summarising results] *is not the result* [expression of cause] of the hydrolysis procedure since Simpson, Neuberger and Liu (1976) described the recovery of Trp as 'excellent'....[backing], but rather *due to* [expression of cause] the interference of carbohydrate in the sulphocylic acid solution [explanatory phenomenon]. (dissertation)

The fourth type posited was same thing reversed: ABDUCTIVE- PHENOMENON. Here the claim consists of a result summary statement of some kind, an expression of implication, and another expression referring to a particular posited - abduced - phenomenon or process implied by the result.

(7) *These estimated intake values* [expression summarising results] *suggest that* [expression of implication] EW lambs maintained similar protein intake 1 wk postweaning but energy intake was decreased by almost half [implied phenomenon]. (article)

(8) *The differences between.... were not always significant....but they were mainly positive* [expression summarising results], *suggesting* [expression of implication] a net transfer of AA from the gut into the blood [implied phenomenon]. (dissertation)

In a fifth type — INDUCTIVE-GENERALISATION (cf. Walton 1989) there is a straightforward datum-claim pattern, where the datum is a result summary statement of some kind, and the claim consists of an expression of implication and a present-tense generalisation:

(9) . ...*results from the present study* [expression summarising results] *indicate* [expression of implication] that feed restriction in pigs *alters* plasma melatonin concentrations .... (article)

(10) *It can be concluded from* [expression of implication] *these results* [expression summarising results] that feeding ground rather than unground MSPB *does not affect* the acidity of the rumen environment..... (dissertation)

A few arguments seemed to be of the type represented by van Eemeren and Grootendorst's 'argument that something is similar to something else' — which

I called ANALOGY. The defining point here was that references were used not as warrants but as data, and results played no part.

(11) ....*based on published results* [data=reference] one cannot conclude with confidence that beta-agonists will consistently improve feed efficiency in growing lambs) (article)

(12) ...higher [molasses] intakes may have been achieved with increased [dry matter] intakes, *as observed by Ruiz (1976), where increased forage intake resulted in increased voluntary intake of molasses.* [data=reference]

A seventh type — DEDUCTIVE —was defined negatively: its members did not fit into any of the above categories. Many of these arguments were in fact mathematical, like the (chained) two in the following sentence:

(13) If endogenous faecal 'crude protein' is subtracted , *then* faecal CP from the feed = 15.2 - 4.5 = 10.7, and *hence* the true digestibility = (7.4 - 10.7)/74........(dissertation)

The rest seemed to be deductive arguments which fitted well into various patterns discussed by argument theorists, like the following:

(14) Insulin concentration may also regulate GH binding to hepatic membrane; *therefore*, decreased insulin concentration would reduce the rate of GH binding to the hepatic receptor. (article: warrant-claim?)

(15) *Since* pulp was molassed before being ground, all four products would have had the same level of addition of molasses. (dissertation: datum-claim?)

While these categories do not exhaust the possible interesting criteria , and are somewhat ad-hoc, they did allow the arguments in the text to be classified. This reflects the peculiar nature of applied-science writing. New findings are cast into familiar forms (*the X was Y, suggesting that A was/is B*) both because the forms have prestige and invest the new findings with it, and simply because they make possible rapid reading of the flood of new findings. More subtly, the form of the scientific article replicates its ideology, and scientists see things in a limited number of categories because their reading presents them in these categories (Berkenkotter and Huckin 1993). Logical forms of argument are defined and restricted by the disciplinary culture and linguistic innovation is avoided for both practical and cultural reasons.

Results

Table 4. Location in the sections of various types of argument

Forms	justify method	status	abductive-cause	abductive-phenomenon	inductive-generalisation	analogy	deductive
First third of text	13	4	15	16	3	3	7
Second third of text	0	7	17	24	4	2	8
Last third of text	2	6	14	24	12	3	4
TOTAL	15	17	46	64	19	8	19

If there is any validity in our classification, one would predict 'justify method' arguments clustering at the beginning of the sections and 'inductive-generalisation' ones at the ends. Table 4 shows that this expectation is fulfilled, and arguments of other types are evenly spread through the text. This comes as no surprise, and agrees with the findings of Brett (1994), Thompson (1993), and others. It does however provide some confirmation of the plausibility of the classification.

Table 5. Arguments classified by source and type

	justify action	status	abductive-cause	abductive-phenomen	inductive-generalisation	analogy	deductive
Dissertations	9	8	28	37	5	3	17
Published articles	6	9	18	27	14	5	2

Table 5 confirms that there are no major differences between the types of argument in dissertations and articles (Swales 1990). There was a relatively low proportion of inductive-general arguments in the dissertations. This is easily explained by the choice of a relatively early part of the Results-Discussion-Conclusions sequence for analysis, and as another predictable result provides a little more confirmation of the validity of the classification. However, some interesting, and unpredicted differences between the two subsamples did appear. 'Deductive' arguments occurred almost exclusively in the dissertations. They may have a knowledge-display function in such contexts, or, since Bazerman (1994) argues that explicit argument tends to be concealed in professional scientific writing, they may be the result of unskilled writing.

Another difference was that most of the 'justify method' arguments in the dissertations had evaluative expressions in their grounds (*necessary, important, unfortunately*) but such expressions were not used in these arguments in the articles (cf. example 1 and 2 for the contrast). This too might be a generic difference, in that the dissertation-writer has, or has to admit, problems which the article-writer does not have, or conceals, or it might be a matter of writer skill.

A third was that far more of the arguments in the dissertations were chained into long sequences (mostly leading up to a methodological decision or an interpretation of the data). Again, dissertation writers may be required to show their reasoning more explicitly or may not know how to conceal it. In any case the finding conflicts with Chou Hare and Fitzsimmons' expectation of more chained arguments in more sophisticated writing. It confirms Swales' (1990) view that dissertations are likely to be more explicit and include more metalanguage, while articles may be shorter and more condensed.

Finally, perhaps because they lack authority, dissertation writers had more hedging in their claims when the marker was already hedged: *suggesting that X may be Y.*

Table 6. Vocabulary in arguments of various types

Forms	justify action	status	abductive-cause	abductive-phenomen	inductive-generalisation	analogy	deductive
Connector:							
verb/noun:	0	0	*	37	13	4	4
(of which, modalised)				(15)	(3)	(3)	0
conjunct(ion)	11	4	2	9	0	3	13
no connector	4	13	44	18	0	1	2
'hedging' on claim content	0	7@	41	9	9	0	0
strengthening on claim content		2@	0				1

* nearly all of these include a modalised phrase of cause/result, etc. *X may be a result of Y*
@ both hedging and strengthening mainly dissertations

The results in Table 6 confirm that some distinction can be made among argument-types in terms of the vocabulary they use. Deductive and problem-solving arguments make frequent use of conjunctions like *since, therefore, if..then, thus*, etc, and are rarely hedged. In this respect they look like the examples of arguments in Walton (1989), Fisher (1988), etc. The argumentative vocabulary in the inductive and abductive arguments is dominated by expressions of implication like *suggest,* and *indicate,* and hedges like *may* and *seem.* This vocabulary is familiar from many investigations (Hopkins and Dudley-Evans, Thompson, Hyland, Myers, etc.) of scientific knowledge claims, but it does not match the vocabulary in the argumentation textbooks. For example, Fisher (1988) lists four modal expressions used to signal reasoning: *must, cannot, impossible, necessarily* where the arguments examined here used *may, possible,* and *probably*. Similarly Toulmin's arguments usually have the claim connected to the grounds by *so, presumably,* and Walton (1989) says that *therefore* marks a conclusion, yet it was rare in inductive and abductive arguments, as were *hence* and *consequently*, and in fact all the phrases listed by Fisher as claim markers. The same applies to Fisher's list of 'reason indicators': *because, for, since, follows from the fact that, the reason being, firstly/secondly, may be inferred from the fact that*.

The explanation may be that the arguments classified here as deductive and problem-solving are actually expository. They (purport to) present agreed knowledge or accepted responses and not to convince of new claims. On the other hand where discussion sections report probabilistic reasoning leading to claims presented as new, they use a specialised vocabulary for it, which involves frequent uncertainty modals and verbs such as *suggest* introducing claims. Possibly the inductive and abductive arguments discussed above show a depersonalised version of the vocabulary used for persuading, while the deductive and problem-solving ones use the vocabulary for explaining. Another way of putting this is that the deductive and problem-solving arguments background any possible doubt, while the inductive and abductive ones foreground it.

The existence of numerous unhedged claims in the texts is mildly unexpected in view of the voluminous literature on hedging in scientific research reports (summarised in Hyland 1996a, b, see also Myers 1989). It is possibly a weakness in the methodology of that literature that it has not been able to compare hedged with unhedged claims, because it has looked for hedges rather than arguments. Nevertheless, because it is probabilistic knowledge claims that are hedged rather than other types, these results tend to confirm Hylands' position that hedging is fundamentally cognitive in function, and used where there is genuine uncertainty, rather than Myers' emphasis on its role in maintaining social relations.

Conclusions

This study has examined a fairly small number of arguments identified in an unsystematic way and classified in what may not be the most illuminating way. But the relatively clear-cut differences found in forms and the conformity of the results with those of previous research suggest that the direction is at least worth pursuing.

It is thus possible to classify the arguments in the results-discussion-conclusions sequences of applied science research in consistent and meaningful ways. The most striking result is that arguments which appear to have different functions vary quite regularly from one another in form. Hedging seems to be restricted to arguments presented as leading to new knowledge claims. Presumably there are a variety of reasons for this. Writers want to represent the genuine logical uncertainty of this type of argument, they want to avoid attack for being too confident, they do not want to appear arrogant, etc.

However, this leads to a paradox. Since scientific writers always hedge claims to new knowledge, the hedge markers become markers of a claim to new knowledge. An unhedged claim using the logical vocabulary *(therefore, must, etc)* without hedging is marked as preparatory or background material. Thus caution becomes a marker of originality, and directness of claim a marker of obviousness. MacDonald (1990) shows how a parallel convention in literary studies can be abused for rhetorical purposes, marking what is obvious as contentious, thus establishing oneself as cautious, and then what is actually contentious as obvious, so that the reader feels impelled to believe the contentious point. It would be interesting to know whether such rhetorical tricks occur in scientific writing. Further investigation of *un*hedged claims might be revealing.

There seems to be more hedging and more classic chained argumentation in dissertations. It is easy to think of possible reasons for this in terms of the functions of the genres, but it is also possible to account for it in terms of writer skill. The fact that most PhD students in experimental science publish articles before they complete their dissertations might make it less likely that the difference is one of skill, but on the other hand it is clear that these articles are extensively co-written by supervisors, so the issue is indeterminate.

Finally, another direction for further research is the use of result summary labels to characterise the results and thus to orient the grounds towards the claim to be made.

References

Berkenkotter, C., and Huckin, T. N. 1993. "Rethinking genre from a sociocognitive perspective". *Written Communication* 10(4): 475-509.

Bizzell, P. 1993. "What is a discourse community?" In P. Bizell (ed), *Academic Discourse and Critical Consciousness*. Pittsburgh, University of Pittsburgh: 222-237.

Brett, P. 1994. "A genre analysis of the results sections of sociology articles". *English for Specific Purposes* 13(1): 47-59.

Chou Hare, V., and Fitzsimmons, D. A. 1991. "The influence of interpretive communities and procedural knowledge in a writing task". *Written Communication* 8(3): 348-378.

Dijk, T. A. van. 1992. "Racism and argumentation: race riot rhetoric in tabloid editorials". In F. H. van Eemeren, R. Grootendorst, A. J. Blair and C. A. Willard (eds), *Argumentation Illustrated* (pp. 243-259). Amsterdam: SICSAT.

Dong, Y. R. 1996. "Learning how to use citations for knowledge transformation: nonnative doctoral students' dissertation writing in science". *Research in the Teaching of English* 30(4): 428-457.

Ducrot, O. 1980. *Les Echelles Argumentatives*. Paris: Les Editions de Minuit.

Eemeren, F. H. van and Grootendorst, R. 1984. *Speech Acts in Argumentative Discussions*. Dordrecht: Foris Publications.

Eemeren, F. H. van and Grootendorst, R. 1992. *Argumentation, Communication, and Fallacies: a pragma-dialectical perspective*. Hilldale, NJ: Laurence Erlbaum Associates.

Fisher, A. 1988. *The Logic of Real Arguments*. Cambridge: CUP.

Francis, G. 1994. "Labelling discourse: an aspect of nominal-group lexical cohesion". In M. Coulthard (ed), *Advances in Written Text Analysis*. London: Routledge.

Hopkins, A. and Dudley-Evans, A. 1988. "A genre-based investigation of the discussion sections in articles and dissertations". *English for Specific Purposes* 7(2): 113-122.

Hyland, K. 1996a. "Talking to the academy: forms of hedging in science research articles". *Written Communication* 13(2): 251-281.

Hyland, K. 1996b. "Writing without conviction? hedging in science research articles". *Applied Linguistics* 17(4): 433-454.

MacDonald, S. P. 1990. "The literary argument and its discursive conventions". In W. Nash (ed), *The Writing Scholar* (pp. 31-62 .)

Martin, J. R. 1988. *Factual Writing: exploring and challenging social reality*. Oxford: Oxford University Press.

Myers, G. 1989. "The pragmatics of politeness in scientific articles". *Applied Linguistics* 10 (1): 1-35.

Myers, G. 1990 . *Writing Biology: texts in the social construction of scientific knowledge*. Madison: University of Wisconsin Press.

Myers, G. 1992. "Textbooks and the sociology of scientific knowledge". *English for Specific Purposes* 11(1) 3-17.

Rees, M. A. van. 1992. "Problem solving and critical discussion". In F. H. van Eemeren, R. Grootendorst, A. J. Blair and C.A. Willard (eds), *Argumentation Illuminated*, 281-291. Amsterdam: SICSAT.

Thompson, D.K. 1993 "Arguing for experimental 'facts' in science: a study of research article results sections in biochemistry". *Written Communication* 10(1): 106-128.

Toulmin, S., Rieke, R. and Janik, A. 1984. *An Introduction to Reasoning*. New York: Macmillan.

Vestergaard, T. (1994). " 'Incomplete' arguments and the principle of relevance". In E.-U. Pinkert (ed), *Universalisme og interkulturel kommunikation*. Aalborg: Aalborg University Press.

Walton, D. N. 1989. *Informal Logic: a handbook for critical argumentation*. Cambridge: CUP.

"Geological Setting/Cadre Géologique" in English and French Petrology Articles: Muted Indications of Explored Places

DACIA F. DRESSEN AND JOHN M. SWALES
The University of Michigan, Ann Arbor

Introduction

There is an established trans-disciplinary tradition of associating genre knowledge with power and authority that at least goes back to Foucault (1972). However, in applied genre analysis the concept of expert manipulation of genres for tactical advantage has only come into prominence since the publication of Bhatia's *Analysing Genre* in 1993. In very recent years, we are thus encouraged to recognize the authoritative, institutional voice which has the "power to use, interpret, exploit and innovate novel generic forms" (Bhatia 1997a: 362). Within this conceptualization, which has many points of contact with Critical Discourse Analysis, this power to innovate is largely seen as being realized through the deliberative mixing or embedding of genres, such as the "commodification" of British university prospectuses (Fairclough 1982), or the increasing uses of promotional strategies in introductions to academic books and similar texts (Bhatia 1995).

While we do not doubt that such hybridizing and commodifying processes can be shown to be at work in contemporary academic and professional genres, we are less sure about the extent of these developments, especially in academic universes of discourse. We suggest that it is just as likely that what we are seeing in contemporary genre studies is a growing realization that genre exemplars are complexly multi-functional; in effect, we are looking as much at new *insights* into discourses as at new textual developments per se. Consider, for example, the case of authors' prefaces to their books since these are discussed as instances of the "new promotionalism" in Bhatia (1997b). In this context we could do well to look once again at the justly famous preface by naturalist John Ray to his *The Wisdom of God manifested in the Works of Creation*, published in 1691:

> In all ages wherein Learning hath Flourished, complaint have been made of the Itch of Writing, and the multitude of worthless Books, wherein importunate Scriblers have pestered the World... I am sensible that this Tractate may likely incur the Censure of a superfluous Piece... *First,* therefore, in excuse of it I plead, That there are in it some Considerations new and untouch'd by others; wherein if I be mistaken, I alledge *Secondly* that manner of Delivery and Expression may be more suitable to some Mens Apprehension, and facile to their understandings. If that will not hold, I pretend *Thirdly,* That all the Particulars contained in this Book, cannot be found in any one Piece known to me, but ly scattered and dispersed in many, so this may serve to relieve those Fastidious Readers, that are not willing to take the Pains to search them out; and possibly, there maybe some whose Ability... will not serve them to purchase... those Books, who may yet spare Money enough to buy so inconsiderable a Trifle.

This marvelously stepped argument indeed makes its four promotional points, but does so with self-conscious irony, with self-deprecatory wit and with highly artful modesty. And, of course, this "so inconsiderable a trifle" is over 300 years old.

While such "puffs" (and their associated demurrals) have probably been around as long as there have been creative written works, there also remain established alternative traditions of diffidence, indirectness and deflection wherein authors tend to decline opportunities for valorizing their own scholarly achievements. There can be patterns of Becker's (1995) "silential relations" here, and these can be shown to be further prone to different levels of modesty in different academic cultures (e.g., Mauranen (1993) for Finnish and Block and Chi (1995) for Chinese). Indeed, matters can be variable within a particular field in a particular culture. In Biology, for instance, the science investigated by discourse analysts *par excellence,* there are worlds of difference in "how much is made of" the discovery of a new mammal, bird, or plant. The first makes the newspaper headlines, the second may warrant the publication of a ten-page article in a leading journal accompanied by color plates (Gross 1990), while the third is only indicated by a short diagnostic paragraph in Botanical Latin and the tiny, unobtrusive abbreviated notation in the front matter — *sp. nov.* (Swales 1998). When a professor of Botany was asked at our institution how many plant species "new to science" were described each year, he replied "I don't know, nor do I suspect does anybody else, but it must be thousands" (W.

Anderson, p.c.). In consequence, here we find muted accounts of discoveries, neither trumpeted in press releases, nor commodified in any other way.

While Biology articles have commanded a great deal of attention (e.g., Myers 1990, Dubois 1987, Selzer 1993), our knowledge of other disciplines, such as Geology, remains scanty. Although Love (1991, 1993) has provided useful analyses of geology textbooks, studies of the Geology research article are currently confined to small-scale investigations into Geotechnical papers (Een 1982, Cox 1995), to Jolivet's (1998) ongoing contrastive work in geology, economics and medicine, and to the general and historical review of the stylistics of geological writing found in Montgomery (1996). Another noticeable gap in the applied linguistics literature is the paucity of studies that contrast French and English scientific writing; for example, French is little mentioned in either Connor (1996) or in the 1997 collection edited by Duszak. Important but rare exceptions are Régent (1985) who argues that differences in visual aspects of the printed page, and in sequences of discursive and communicative acts, are caused by slightly different philosophies of science in the two cultures, and again, Montgomery (1996) who discusses the more "literary" and "writerly" qualities still to be found in Francophone geological papers, but now largely absent from their Anglophone counterparts.

In this chapter, we attempt to respond to these lacunae by investigating a corpus of 20 texts in the important geological sub-discipline of petrology. As its name implies, petrology is concerned with the description, classification, and analysis of rocks. We have noted that a certain number of petrology articles contain a subsection intervening between the introduction and the methodology/results sections. This subsection, most typically called "Geological Setting" (GS) in English and "Cadre Géologique" in French, has not to date been reported on in the now-extensive literature on the schematic structuring of subsections in the scientific research article. We argue that this addition reflects an alternate framing for geological research (i.e., one that originates in phenomena observable on the terrain rather than around laboratory analyses). As such, it is primarily characterized by a series of general-to-specific descriptive statements which describe the topographical, historical, and physical features of the terrain from which the rock samples have been taken. The depiction of this type of background information has, over the years and decades, become largely conventionalized in the form of a part-genre (Ayers 1994).

On one plane, then, these GS elements provide a historical and topographical contextualization for the analytic work to be discussed in the body of the paper. However, on another plane these elements also serve to establish the authors' credentials as experts in their chosen and often remote locales, although today this credentialization is only rarely achieved by "self-promotional" human agency, but via muted and indirect means. We also show in the chapter that the French and English texts in the corpus show surprising similarities at (macro) discoursal strategies, but diverge consistently in grammatical tactics.

Background

Our first look at petrology articles showed that in the first 20 articles, 12 (60%) contained a sub-section that we eventually decided to be the GS part-genre. We then scanned further petrology periodicals to arrive at a working corpus of 10 articles in English and 10 in French. We estimate that these sections are likely to occur about 50% of the time in petrology. Details of the corpus are given at the end of this chapter[1]. The actual articles are listed in the Appendix and are referred to by letter (A-T) throughout the remainder of this chapter.

The average length of the GS section was 619 words for English and 659 words for French, with outliers at 298/237 and 1453/1130 for English and French, respectively. In 19 cases, the GS section was initially marked by a named subheading. The nomenclature here is fairly, but not entirely, consistent. The use of "geological/géologique" (in 15 of 20 titles), of a more general denomination than "lithographical" or "geochemical", combined with "setting/ cadre/contexte" (in 14 of 20 titles) effectively acts as a discoursal indication that the GS contains general, background information. A few cases, however, do not adopt this seemingly more conventional naming practice; one is untitled, one uses "field relations, and the remaining two indicate the precise subject of the GS (e.g., "Granitoids in SW Japan" and "Les granitoïdes des dolérites"). In those cases where a typographically marked onset was not clearly indicated, we looked for the "discoursal patterning" to be discussed shortly, and in the rather more frequent occurrence of unmarked closure of the GS section, we set the ending of the section at the juncture where authors switched to geochemical analyses and quantified assays of the rocks. In addition, we have been assisted in this research by a number of Anglophone and Francophone geologists, who have generously offered their time and expertise as specialist informants. These individuals are named in the Acknowledgments.

Features of Geological Setting sections

Discoursal patterns

If the GS is to be taken as a part-genre, we can assume that its exemplars would reflect disciplinary concerns specific to Geology. In both the French and English sub-corpora, elements reflective of these concerns were indeed found to occur with a high degree of frequency very early on in the GS, normally within the first paragraph. The sub-moves of these "GS openings" are noted as follows: Localization; Nomenclature; Visual support; Formation composition; Age; Geological activity; Prior literature; Sublocalization; and Brief description of rocks.

Occurring in all twenty texts, "Localization" may simply indicate the name of a geographical place (e.g., "the Chilwa Island carbonatite"), or may be determined in terms of the geological terrain (e.g., The Karakoram batholith runs the length of the High Karakoram range in northern Pakistan, Ladakh, and westernmost Tibet). In the next most frequent type of sub-move (i.e., "Nomenclature", in 19 out of 20 texts), the rock- or formation-type is identified, as in "The Laramie *anorthosite* complex" or "les *dolérites*"; it often occurs in close conjuncture with the given location. We have also noticed in 17 of the 20 texts that authors provide some map or figure to help the reader visually determine the exact geographical location of the formation, as well as how it appears on the terrain in relation to surrounding geological features. Equally frequent is the sub-move which identifies the formation's composition, in terms of its rocks and/or minerals. Very often, this sub-move directly proceeds a detailed petrological description of the minerals and their composition. A majority of texts in our corpus (12 of 20) also indicate both the geological age of the formation, as well as what type of geological activity has taken place (e.g., the chemical or compositional alterations of surrounding elements). These patterns can be seen in the following three examples:

(1) Localization   In the Oatlands district,
     Visual support   (fig.1)
     Nomenclature   basalts of the
     Age   Oligocene and Early Miocene age
     Geological activity   occur as isolated flow remnants and plugs. They form part of a much larger Cainozoic basalt
     Sublocalization   province that extends along the length of eastern Australia.
     Formation composition   Upper crustal rocks include carbonaceous shales and quartz sandstones... [A]

(2)	Nomenclature	The Laramie anorthosite complex (LAC)
	Localization	is exposed over 800 km2 in southwestern Wyoming
	Visual support	(fig. 1)
	Age	and represents the northernmost component of wide-spread 1.4 Ga magnetism in the western USA.
	Geological activity	The LAC is inferred to have intruded the Cheyenne belt,
	Sublocalization	a major terrane boundary that separates Archean rocks of the Wyoming province to the north from accreted 1.7-1.8 Ga rocks of the Colorado province to the south. [E]
(3)	Nomenclature	*Le massif volcanique du Mbam*
	Localization	*occupe une superficie de 450 km² sur la bordure NW du plateau Bamoun, unité médiane des Hauts Plateaux de l'Ouest-Cameroun.*
	Formation composition	*Sur un substratum constitué de granitoïdes*
	Age	*d'âge probablement panafricain,*
	Geological Activity	*il est constitué de deux épisodes volcaniques distincts*
	Visual support	*(fig. 1).* [T]

"The Mbam volcanic massif crops an area of 450 km² on the NW range of the Bamoun Plateau, the central zone of the Hauts Plateaux of western Cameroon. Its basement consists of granitoids, probably of Panafrican age, and the massif consists of two distinct volcanic episodes."

An immediately obvious feature of these GS openings (and of others in the corpus) is the amount of territory covered in only 2-3 sentences. They are further marked by a movement from general to specific, and by their handling of the Setting in a tight, highly conventionalized, and largely uniform way.

Lexical verb choice

In order to capture another aspect of these texts, all lexical verb choices (both finite and non-finite) in the twenty English and French GS sections were counted. Around 440 verbs were found for each sub-corpus, and were situated in four types of rhetorical context: (1) the description of the geological setting itself; (2) accounts of previous studies of the area; (3) accounts of what the authors have done themselves (i.e., reporting verbs as in "We found..."); and (4) the authors' metadiscourse (e.g., "Details are provided in Appendix A"). The very clear predominance of the descriptive task in both English and French (around 88% and 95% of all verbs, respectively) reinforces the textual impression that the authors here are dedicated to providing the broader geological context for their study. References to previous studies (category 2) account for fewer than 10% of the verbs, while categories 3 and 4 have very few tokens. Table 1 reflects the kinds of lexical verbs which occur throughout the corpus.

Table 1. Lexical verbs occuring four times or more

ENGLISH	FRENCH
83 -- be	64 -- être
25 -- occur	15 -- constituer
14 -- form, intrude	12 -- avoir
13 -- have	11 -- associer, contenir
12 -- contain	10 -- correspondre
11 -- consist of, show	8 -- présenter (se)
10 -- cut, indicate	7 -- caractériser, recouper, situer (se)
7 -- separate, expose	6 -- affecter, s'agir de, limiter (se), montrer
6 -- include, represent	5 -- apparaître, appartenir, comporter, comprendre, représenter, varier
5 -- accompany, associate, range	4 -- apporter (se), conduire, développer (se), distinguer, exclure, expliquer (se), interpréter, observer, provenir, reposer, retenir, séparer, suggérer, suivre, supposer
4 -- appear, define, locate	

As the most significant category, descriptive lexical choices have their own story to tell. First, the high frequencies of verbs such as *be, occur, have, contain, consist of, show* or *include* in English and *être, constituer, avoir, associer, contenir, correspondre* or *se présenter* in French underscore some preference for a stative synchronic account rather than an unfolding diachronic account of geological activity. This impression is corroborated by the fact that the majority of the verbs are stative (around 60% for both French and English) and intransitive (around 40%).

Another aspect emerging from the initial count of lexical verb choices is that few of the verbs in either language are explicitly geological. *Emplace* (3 occurrences) in English is certainly one, but most are instead "sub-technical" (cf. Birch 1996; Lowe 1995). There are, of course, some special geological uses — and meanings — of verbs of more general use. The most prominent of these in English are *intrude, cut, expose,* and *associate,* while in French they are *associer* and *recouper.* Finally, to round out the picture, we list those few instances in the corpus of lexical verb choices that are truly geological in nature: *cross-cut, accrete, grade, root,* and *transect;* or in French: *dater, effectuer, mettre en place, affleurer, alimenter,* and *étager.*

Through the choice of these verbs, petrologists, rather than describe their own interactions with their research objects, are seen to describe the actions and states of the rocks, minerals and formations which they have found either during their own terrain observations, or as described in the literature. Thus, whereas the tendency in some earlier articles may have been to engage in "traveler's tales", relating the arduous nature of field work and of discovery in off-the-map kinds of places, by this decade the GS section is largely detached from human agency. The agentive narratives common in earlier petrology GS sections have disappeared except for relatively infrequent and mostly "weak-author-oriented" (Swales 1990) references to previous work on the research site.

Tense choices

The tense data shown in Table 2 further corroborate the unfolding story of the GS part-genre. As might now be expected, we find in both languages a strong predominance of present-tense description. Beyond that, there are some interesting variations between French and English.

Table 2. Tense choice correspondences with rhetorical category

Rhetorical Categories	Description	Emplacement	Commentary	Previous studies	Present work	Metadiscourse
ENGLISH 347 total verbs	252 (73%)	30 (9%)	26 (7%)	26 (7%)	7 (2%)	6 (2%)
Present	97%	-	81%	12%	42%	83%
Past	1%	93%	4%	42%	29%	-
Perfect	2%	7%	5%	46%	29%	-
Future	-	-	-	-	-	17%
FRENCH 389 total verbs	275 (71%)	40 (10%)	44 (11%)	13 (3%)	10 (3%)	7 (2%)
Présent	98%	50%	100%	31%	-	67%
Passé Composé	2%	45%	-	69%	100%	-
Imparfait	-	5%	-	-	-	-
Futur	-	-	-	-	-	33%

The depiction of rock and formation emplacement in English, for example, shows the diachronic unfolding of geological time, using the past and perfect tenses.

> 4. Many of the sills <u>have separated</u> into layered granite-pegmatite couplets (Rockhold et al., 1987, Duke et al., 1988, 1992). [D].
>
> 5. These dykes <u>were emplaced</u> after the main period of penetrative deformation but still during granulite facies metamorphism [which] <u>occurred</u> in the time span between 2.53 and 2.48 b.y. and closely <u>followed</u> the formation of the igneous batholiths (Buhl, 1987, Vidal et al., 1988, Peucat et al., 1989). [B]

Here French differs markedly from English, in that around half of the "emplacement" type verbs actually occur in the present tense, and that an extended account of geological activity is often "tucked away" into non-primary grammatical elements, such as a noun phrase or the past participle.

> 6. *Les leucogranites peralumineux carbonifères <u>se mettent en place</u> le long de la zone de collision et <u>forment</u> des alignements très caractéristiques suivant de grandes failles...* [L]
>
> "The carboniferous peraluminous leucogranites are emplaced along a collision and form characteristic alignments along major faults..."
>
> 7. *Les leucogranites de la Brâme seraient synchrones de la fin des événements tectono-métamorphiques acadiens dévéloppés entre 400 et 360 Ma dans le nord-ouest du Massif central.* [K]
>
> "It would appear that the Brâme leucogranite is synchronous with the end of the tectono-metamorphic Acadian phase which occurred between 400 and 360 Ma in the northwest Massif Central."

As we can see in example 7, the present conditional (seraient) anchors the emplacement process, seen here as a carefully constructed unit consisting of a temporal marker (adjective "synchrones"), combined with a temporally-bounded noun phrase ("la fin des événements tectono-métamorphiques acadiens") with a corresponding past participle ("développées"), and an indication of the time period ("entre 400 et 360 Ma"). The above example is by no means a solitary case; in fact, this type of multi-faceted strategy for indicating the emplacement process occurs throughout the French GS sections in our corpus, an observation confirmed by our French native-speaker geology informant.

As can also be seen in Table 2, authorial comment occurs overwhelmingly in the present tense in both languages, and is accompanied by linguistic markers of hedging, such as modals and epistemic verbs, as well as by references to past work. Most occurrences in this category are for the purpose of evaluation, as seen in the following examples.

1. <u>Si l'on peut exclure</u> leur origine par démantèlement de niveaux sédimentaires indurés, <u>on ne peut pas exclure</u> un bref transport en milieux aqueux, les centres émissifs situés probablement en mer à l'ESE de l'île de Kasserine (Sassi, 1974). [O]

   "Although it is possible to rule out an origin from the breakdown of hardened sedimentary strata, one cannot rule out a short-time transport in an aqueous medium, the vents being probably located under the sea to the ESE of Kasserine Island (Sassi, 1974)."

2. Un seul processus, continue, de genèse et de mise en place fondé sur la continuité structurale et géochimique entre le complexe de la Brâme et le massif de Saint-Sylvestre <u>est envisagé</u> par de nombreux auteurs (...). D'autres <u>se sont appuyés</u> sur des données radiométriques (...) et sur une étude de pétrologie structurale (...) pour proposer une formation en plusieurs étapes. [K]

   "A single and steady process for the genesis and emplacement based on the structural and geochemical continuity between the Brâme complex and the Saint-Sylvestre massif is envisioned by numerous authors (...). On the basis of radiometric data (...) and a structural petrology study (...), other authors have argued for a multi-stage formation."

3. Such large rigid-body rotations <u>could entirely invalidate</u> the use of ductile movement directions, measured in a present-day coordinate frame, for the reconstruction of palaeotectonic environments (e.g., Vauchez & Nicholas, 1991). [F]

One final note about the tense data concerns the authors' accounts of previous studies. In general, the GS in the present corpus gives the appearance of being a neutral terrain, where the authors are especially concerned with establishing what is known about the formation. This observation is corroborated by a number of features. First, we can note the overwhelmingly "weak-author-orientation" (e.g., Swales 1990) of the citations present in the corpus. Furthermore, the great majority of references to past work adopt a parenthetical strategy, which in part explains the small percentage of verbs dedicated to this end. Moreover, authors cite themselves infrequently; only three of an average of twenty citations per article refer to their own previous work. In this sense, then, the GS does not seem to be considered the right place to stake out territory nor to establish the researchers' credibility and authority by reporting on the extensiveness of their previous work on the site.

Noun phrases as markers of professional expertise

As we have already noted, a demonstration of extensive fieldwork in GS sections is hardly typical. Nonetheless, the authors must still signal, however mutedly, that they have indeed been on the terrain. Expertise in the locale and the authoritative voice of the expert are thus to be demonstrated by the description of the data. It is here that the linguistic and discoursal markers which proclaim to the reader "We were there" are to be found, rather than in the type of agentive narrative common in earlier petrology texts.

As seen previously, verbs in GS sections tend to avoid the technicalities of geological descriptions, which in both English and French are shunted into instrumental noun phrases. They both may also further rely on non-finite verbal (participial) phrases or verbal modifiers to carry the technical aspects of the account.

(1) The <u>intrusions</u> have domal structures which grew laterally <u>by continual emplacement</u> of numerous sills and dikes, <u>indicating</u> extraction on small batches of melts from the sources. [D]

In English, the extensive use of extended (and essentially processual) noun phrases and non-finite verbal, adverbial and adjectival modifiers as "product-markers", are densely interspersed throughout the GS section, and describe the physical development of the terrain (i.e., the "process"). This is clearly demonstrated in the following example:

(2) The peridotite overlies high-grade gneisses and marbles...<u>along</u> an essentially low-angle brittle <u>thrust</u> marked by extensive <u>brecciation</u> discernible <u>over</u> a distance of up to 100 m <u>away from</u> the context. [F]

We might note that the two preceding passages are marked especially by process nouns used to describe some physical feature of the rock or formation (e.g., emplacement, extraction, thrust, brecciation), as well as by adverbial movement (e.g., along a thrust, over a distance, up to 100 m away from), resulting in a complex and skillful description made by the geologist on the terrain. Because convention apparently no longer allows geologists to come out and say "<u>We</u> went to the site, <u>we</u> picked up some rocks, and <u>we</u> saw that they were situated in a certain manner, which suggests to us that....", this type of construction serves as a notice to the readers that there has been an actual "taking in" of the terrain with a specialist's eye, rather than an account which has merely been gleamed from the literature.

Such extended phrasal description is extremely common throughout the English sub-corpus, where nearly every sentence uses this compact strategy for relaying the technicalities of geological description. In this sense, then, we can see that the directness of the English texts is best characterized by a reduction to bare verbal movement, where one can see a "skeletal" descriptive subject-verb embedded in the technicalities of geological description.

The placement of technical jargon into noun phrases is also seen to occur in French, although with some slight differences.

> (3) *Ils contiennent deux types d'enclaves, (i) des enclaves microgrenues sombres suggérant l'intervention précoce de magmas basiques avec <u>possibilité d'hybridation</u> par des <u>processus de type assimilation-cristallisation fractionnée</u> et (ii) des enclaves crustales de socle métamorphique et de métasediments, intérprétées comme étant des restites d'une source essentiellement crustrale.* [L]

> "[In these rocks], two types of enclaves are present, (i) microgranular dark enclaves suggesting an early intervention of basic magmas with possible hybridization through assimilation-fractional crystallization processes and (ii) crustal enclaves either of metamorphosed basement or of metasediments, interpreted as restitic material from an essentially crustal source."

In particular, we can note that the NPs which contain the account's "geological technicalities" remain more abstract than in English, where the more processual NP describes a physical state currently detectable on the terrain. While processual NPs are also present here (e.g., hybridation, assimilation-cristallisation fractionnée), they are in fact embedded within a larger, abstract noun phrase (e.g., possibilité, processus). As we will see shortly, French GS sections are marked by both more abstract and process NPs than is English.

Choice of voice and subject type: The agency of "rocks"

As we have remarked earlier, GS sections are also marked by geologists' need to reserve agent-centrality for the rock itself. The central and even unusual role given to rocks and formations can be seen in the following passage:

> "The Karakoram terrane, along the northwest frontiers of Pakistan and India, forms the southern continental margin of the Asian plate (Desio 1964). It lies immediately north of the Tethyan suture zones which mark the zone of collision between India and Asia (Fig. 1). The Shyok suture zone (SSZ) separates

the Kohistan arc-microplate from the Karakoram terrane in the north and the Main Mantle Thrust (MMT) places the Kohistan arc-microplate southwards over upper and mid-crustal rocks of the Indian plate. Sedimentology along the Indus suture zone (ISZ) and north Indian plate margin in Ladakh and south Tibet suggests that closure of Tethys along the ISZ, and collision of India and Asia occurred between the early and mid-Eocene at ca. 50 Ma..." [H]

Immediately striking in this passage is the total absence of human researchers, where, as suggested by an abundance of inanimate subjects with active verbs, rocks and formations appear to act independently of the human hand, despite the obvious necessity of such human activities as geological mapping.

While it is widely believed that the passive is the key grammatical marker for attaining such humanless prose in English "scientific" styles, in contrast with French (Vinay and Darbelnet 1996), analysts have in fact shown that the occurrence of the passive voice in English scientific prose is not as high as once thought. Indeed, several studies have shown the English language to have a wide range of strategies available to remove "the human agent", rhetorically allowing for seemingly greater objectivity (e.g., Tarone et al. 1981, Master 1991). Certainly it seems clear from the above passage that the passive is quite absent, and that the centrality of rocks and formations is indicated by other means.

Of 347 possible subject nouns in English and 389 in French, nouns referring to rocks and formations are by far the most frequent subject type found in both the French and English sub-corpora (77% and 86% of all subject NPs, respectively), clearly highlighting the essential role that they play in GS sections. Moreover, our analysis of the interplay of subject nouns and voice in these sections shows that rock and formation subject nouns in both the English and French sub-corpora are most likely to occur with the relational (26% and 27% respectively), as illustrated in the following examples.

(4) *Du quartz interstitiel ou des intercroissances quartz-feldspath <u>sont souvent présents</u> en faible quantité.* [P]

"Interstitial quartz or quartz-feldspath intergrowths are often present in small amounts."

(5) The LAC <u>represents</u> the northernmost component of widespread 1.4 Ga magnetism in the western USA. [E]

Both French and English further demonstrate a preference for using the transitive with this type of noun rather than the passive (20% transitive vs. 16% passive in French, and 25% transitive vs. 19% passive in English).

(1) *Les granitoïdes calco-alcalins carbonifères <u>recouvrent</u> les champs de granite d'arc, des granites de collision, et des granites syn à post-collision.* [L]

"Calcalkaline carboniferous granites encompass the fields of arc granites, collisional granites and syn- to post-collisional granites."

(2) *Le bassin de Maknassy-Mezzouna <u>englobe</u> une succession de petits bassins.* [O]

"The Maknassy-Mezzouna basin covers a series of smaller basins."

(3) The massif <u>is transected by</u> sinistral strike-slip faults with estimated displacements of up to several hundred metres. [F]

As we have further noticed, the overall use of the passive voice in English GS sections is somewhat higher than in French (27% and 20% respectively), thereby lending credence to suggestions that Francophone writers need to pay attention to the English passive in scientific writing (e.g., Vinay and Darbelnet 1996, Villez 1996, Planes 1996). Although it uses a slightly higher number of passives than French, however, it is in fact in English that rock and formation NPs have the strongest tendency to occur with the transitive, and not in French, as we might have expected given that French uses more transitives overall (31% vs. 26% in English).

The trends of voice with abstract and process subject nouns, on the other hand, are quite different. First, abstract and process noun subjects occur nearly twice as frequently in French (20%) as they do in English (11%). Furthermore, when we focus on the correspondences between the choice of subject NP and voice, this preference in fact points to a significant contrast between French and English writing styles, as is indicated in Table 3.

Table 3. Process and abstract nouns correlated with choice of voice

	Transitive	Stative/ Intransitive	Relational	Passive
English	10%	-	10%	80%
French	39%	24%	20%	17%

As we can see, English GS sections show a very clear and significant preference for using the passive voice with this type of subject NP.

(6) <u>Samples for study</u> <u>were collected</u> from five localities. <u>Descriptions</u> of these and of the collected samples <u>are given</u> as follows. [A]

In contrast, while French authors do not show as clearly a marked preference for one particular voice as their Anglophone counterparts, they are nonetheless most likely to refer to abstract and process nouns in the transitive voice, and least likely to refer to them in the passive.

(7) *<u>Les rapports isotopiques</u> initiaux du strontium élévés <u>indiquent</u> une source magmatique crustale.* [L]

"High initial strontium isotopic ratios indicate a crustal magmatic source."

(8) *<u>Une coupe simplifée</u> (fig. 3) <u>résume</u> la série lithologique de l'Oued Abiod.* [O]

"A simplified cross-section summarizes the lithologic succession of Oued Abiod."

The overall tendencies found in this analysis thus lead to the observation that the writing style of French authors in GS sections is more process-oriented, abstract and deterministic than that found in the English sub-corpus. Furthermore, while we most assuredly do not have the impression from these texts that passive use by Anglophone writers is an indication of their being less "direct" in their description of the geological terrain, nor does the slightly higher use of passives in English in any way demonstrate that they have more difficulty in formulating firm judgments as some have suggested (e.g., Vinay and Darbelnet 1996), their texts are nonetheless clearly marked by a different descriptive strategy. In contrast, English GS sections appear more "boiled-down" and even at times clinical than those texts marked by the more "process-oriented" descriptive strategies of the French writer.

Discussion

Here, then, are texts which are rich complexes of description, geological process, the evaluation of prior claims, and the establishment of professional authority. As we have seen, they are linguistically quite similar on a macro-discoursal level, as seen in the manipulation of sub-moves in GS openings, the types of rhetorical categories governing verb usage, and the purposefully descriptive and objective distancing of the researcher which allows the rocks to "tell their own story". We have also seen important variations on a more local level. English GS sections, for example, while making the rock and formation the central active agents in its prose, minimizes the input of the author-researcher by overwhelmingly marking researcher and intellectual activity with the passive voice. Quite to the contrary of French, where it is more common to embed all activity, geological technicalities as well as researcher evaluation in active processual and abstract expression. These observed differences may provide us with one indication of why French geological texts may leave the impression of being more "writerly" and "literary" than the more "clinical" and "monotonic" English texts (e.g., Montgomery 1996).

These micro-level differences also confirm the results of other cross-linguistic genre studies (e.g., Melander et al. 1997), where reported differences at a local level are explained by the particularities of the local discourse communities in question. To our knowledge, only a handful of researchers have undertaken the task of exploring some of these differences in French and English (e.g., Régent 1985, Donahue 1998). To some extent, the characteristics particular to the Francophone and Anglophone discourse communities in Geology, although not highlighted in this chapter, have begun to emerge through our discussions with the geology informants consulted for this study.

Finally, we have been at great pains in this chapter to stress the petrology-intrinsic motivation for the existence of the additional GS part-genre and for its transitional placement between the "true" introduction and the petrological analyses. We have also hinted at its discoursal evolution in recent decades from reports of geological *expeditions* to contemporary succinct and cumulative descriptions wherein the rocks and the geological processes that impinge upon them tell their own stories. It has, of course, not escaped our notice that petrology is not the only field-based discipline; others that easily come to mind include much of archeology, anthropology, ecology, fieldwork in biology, and even studies of endangered languages in linguistics or of isolated speech communities in sociolinguistics. While we might expect all researchers in field-based disciplines to establish their authority and credibility as experts in their chosen research-sites by various contextual and textual strategies, it remains, for now, an open question as to whether their tactics will be similar or different to the Anglophone and Francophone petrologists we have examined in this study.

Acknowledgments

We would like to thank our geology informants Neil Irvine, Hatten S. Yoder Jr., James Farquhar, Dan Frost, Dave Teter, Louis Latouche, Gabriel Gohau, Hamouda Hemissi, as well as the Geology Department at the Université de Tunis I for their time, interest, and insightful comments. We would especially like to thank Tahar Hammouda for the phenomenal amount of experiential insight, time and support he has brought to the study.

Notes

[1] The corpus consists of twenty Geological Setting sections from four different French and British Petrology and Earth Science journals from the years 1990-1996. Six articles were taken from the *Journal of Petrology* and four from the *Transactions of the Royal Society of Edinburgh*. Five articles each were also taken from the *Bulletin de la Société Géologique de France* and from the *Comptes Rendus de l'Académie des Sciences: Sciences de la Terre et des Planètes*.

Appendix

English articles consulted

A. Adam, J. 1990. The geochemistry and experimental petrology of sodic alkaline basalts from Oatlands, Tasmania. *Journal of Petrology*, 31: 1201-1224.

B. Srikantappa, C., Raith, M., Touret, J. 1992. Synmetamorphic high-density carbonic fluids in the lower crust: Evidence from the Nilgiri Granulites, Southern India. *Journal of Petrology*, 33: 733-760.

C. Simonetti, A., Bell, K. 1994. Isotopic and geochemical investigation of the Chilwa Island carbonatite complex, Malawi: Evidence for a depleted mantle source region, liquid immiscibility, and open-system behaviour. *Journal of Petrology*, 35: 1597-1621.

D. Nabelek, P., Glascock, M. 1995. REE-depleted leucogranites, Black Hills, South Dakota: A consequence of disequilibrium melting of monazite-bearing schists. *Journal of Petrology*, 36: 1055-1071.

E. Mitchell, J., Scoates, J., Frost, C., Kolker, A. 1996. The geochemical evolution of anorthosite residual magmas in the Laramie Anorthosite Complex, Wyoming. *Journal of Petrology*, 37: 637-660.

F. Van der Wal, D., Vissers, R. 1996. Structural petrology of the Ronda peridotite, SW Spain: Deformation history. *Journal of Petrology*, 37: 23-44.

G. Hald, N., Waagstein, R. 1991. The dykes and sills of the Early Tertiary Faeroe Island basalt plateau. *Transactions of the Royal Society of Edinburgh*, 82: 373-388.

H. Searle, M., Crawford, M., Rex, A. 1992. Field relations, geochemistry, origin and emplacement of the Baltoro granite, Central Karakoram. *Transactions of the Royal Society of Edinburgh*, 83: 519-538.

I. Krogstad, E., Walker, R. 1996. Evidence of heterogeneous crustal sources: The Harney Peak Granite, South Dakota, U.S.A. *Transactions of the Royal Society of Edinburgh*, 315: 331-337.

J. Nakajima, T. 1996. Cretaceous granitoids in SW Japan and their bearing on the crust-forming process in the eastern Eurasian margin. *Transactions of the Royal Society of Edinburgh*, 315: 183-191.

French articles consulted

K. Léger, J.-M., Wang, X., LaMeyre, J. 1990. Les leucogranites de Saint-Goussaud en Limousin: pétrographie, éléments majeurs et traces dans le sondage de Villechabrolle (projet Energeroc). *Bulletin de la Société Géologique de France*, 6: 515-524.

L. Lagarde, J.-L., Capdevila, R., Fourcade, S. 1992. Granites de collision continentale: l'exemple des granitoïdes carbonifères dans la chaîne hercynienne ouest-européenne. *Bulletin de la Société Géologique de France*, 163: 597-610.

M. Vicat, J.-P., Pouclet, A. 1995. Nature du magmatisme lié à une extension pré-panafricaine : les dolérites des bassins de Comba et de Sembé-Ouesso (Congo). *Bulletin de la Société Géologique de France*, 166: 355-364.

N. Lahaye, Y., Blais, S., Aubray, B., Ruffet, B. 1995. Le volcanisme fissural paléozoïque du domaine nord-armoricain. *Bulletin de la Société Géologique de France*, 166: 601-612.

O. Béji-Sassi, A., Ouazaa, L., Clocchiatti, R. 1996. Les inclusions vitreuses des ilménites, apatites et quartz des sédiments phosphatés de Tunisie : témoignages d'un volcanisme alcalin d'âge paléocène supérieur à éocène. *Bulletin de la Société Géologique de France*, 167: 227-234.

P. Roach, R., Lees, G., Rowbotham, G. 1992. Le champ filonien paléozoïque du Trégor, Nord du massif armorician. *Comptes Rendus de l'Académie des Sciences*, 315: 813-820.

Q. Sagon, J.-P., Sabourdy, G. 1993. Le xénotime, un marqueur de l'unité inférieure des gneiss dans le centre Limousin, Massif Central français. *Comptes Rendus de l'Académie des Sciences*, 317: 1461-1468.

R. Quémeneur, J., Lagache, M., Correia Neves, J. 1993. La pegmatite Urubu, Araçuai, Minas Gerais (Brésil), exemple de pegmatite complexe à pétalite : zonalité minéralogique et géochimie des micas et tourmalines. *Comptes Rendus de l'Académie des Sciences*, 317: 1425-1432.

S. Gasquet, D., Fernandez, A., Mahé, C., Boullier, A.-M. 1995. Origine des rubanements dans les granitoïdes : exemple du monzogranite de Brignogan. Plouescat (NW du Massif armoricain). *Comptes Rendus de l'Académie des Sciences*, 321: 369-376.

T. Moundi, A., Menard, J.-J., Reusser, E., Tchana, F., Dietrich, V. 1996. Découverte de basaltes transitionnels dans le secteur continental de la Ligne du Cameroun (Massif du Mbam, Ouest-Cameroun). *Comptes Rendus de l'Académie des Sciences*, 322: 831-837.

References

Ayers, G. 1994. *Are abstracts changing? A preliminary investigation through the analysis of the short texts accompanying the articles in Nature.* Unpublished M.A. Dissertation, University of Birmingham, U.K.

Birch, S. 1996. "French researchers publishing in English: An analysis of a corpus of first drafts." *Anglais de Spécialité* 11-14: 75-88.

Becker, A.L. 1995. *Beyond Translation.* Ann Arbor, MI: University of Michigan Press.

Bhatia, V.K. 1993. *Analysing Genre: Language Use in Professional Settings.* New York: Longman.

Bhatia, V.K. 1995. "Genre-mixing in professional communication — the case of private intentions v. socially recognized purposes." In P. Bruthiaux, T. Boswood and B. Du-Babcock (eds), *Explorations in English for Professional Communication.* City University of Hong Kong, 1-19.

Bhatia, V.K. 1997a. "The power and politics of genre." *World Englishes* 16: 359-371.

Bhatia, V.K. 1997b. "Genre-mixing in academic introductions." *English for Specific Purposes* 16: 181-195.

Block, J. and Chi, L. 1995. "Comparison of citations in Chinese and English academic discourse." In D. Belcher and G. Braine (eds), *Academic Writing in a Second Language: Essays on Research and Pedagogy.* Norwood, NJ: Ablex, 231-274.

Connor, U. 1996. *Contrastive Rhetoric: Cross-Cultural Aspects of Second-Language Writing.* New York: Cambridge University Press.

Cox, J. 1995. "Analysing geotechnical engineering abstracts: Towards a pedagogical template". *ESP Malaysia* 3: 136-144.

Donahue, C. 1998. *A Cross-Cultural Linguistic/Textual Analysis of Student Writing at the Pre-University and University Levels.* Unpublished Ph.D. Dissertation, Université de Paris V, France/Northeastern University, US.

Dubois, B.L. 1987. "Something on the order of around forty to forty-four: Imprecise numerical expressions in biomedical slide talks." *Language in Society* 16: 527-541.

Een, J. 1982. "Tense usage in the reporting of past research in geotechnical writing." In *Working Papers in ESL, Vol. 2,* 72-91. Minnesota Working Papers in ESL: University of Minnesota.

Fairclough, N. 1982. *Discourse and Social Change.* Cambridge: Polity Press.

Foucault, M. 1972. *The Archaeology of Knowledge.* New York: Pantheon Books.

Gross, A.G. 1990. *The Rhetoric of Science.* Cambridge, MA: Harvard University Press.

Irvine, T.N. and Rumble, D. 1992. *Journal of Petrology: A Writing Guide for Petrological (and Other Geological) Manuscripts.* Oxford: Oxford University Press.

Jolivet, E. 1998. *La Communication Scientifique Orale. Etude des Caractéristiques Linguistiques et Discursives d'un "Genre". Application à Trois Disciplines: Géologie, Médecine, Physique.* Unpublished Ph.D. Dissertation. Université de Bordeaux II, France.

Love, A. 1991. "Process and product in geology: An investigation of some discourse features of two introductory textbooks." *English for Specific Purposes* 10: 89-109.

Love, A. 1993. "Lexico-grammatical features of geology textbooks: Process and process revisited." *English for Specific Purposes* 12: 197-218.

Lowe, I. 1996. "Non-verbal devices in pre-university science: The extent of correspondence between English and French." *English for Specific Purposes* 15(3): 217-232.

Master, P. 1991. "Active verbs with inanimate subjects in scientific prose." *English for Specific Purposes* 10: 15-33.

Melander, B., Swales, J.M. and Fredrickson, K.M. 1997. "Journal abstracts from three academic fields in the United States and Sweden: National or disciplinary proclivities." In A. Duszak (ed), *Culture and Styles of Academic Discourse*. Berlin: Mouton de Gruyter, 251-272.

Mauranen, A. 1993. *Cultural Differences in Academic Rhetoric*. Frankfurt: Peter Lang.

Montgomery, S. 1996. *The Scientific Voice*. London: Guilford Press.

Myers, G. 1990. *Writing biology: Texts in the Construction of Scientific Knowledge*. Madison: University of Wisconsin Press.

Planes, L.-M. 1996. "Activer la voix passive chez les apprenants et professionnels du milieu aéronautique." *Anglais de Spécialité* 11-14: 459-465.

Régent, O. 1985. "A comparative approach to the learning of specialized written discourse." In P. Riley (ed), *Discourse and Learning*. London: Longman.

Selzer, J. (ed). 1993. *Understanding Scientific Prose*. Madison, WI: University of Wisconsin Press.

Swales, J.M. 1990. *Genre analysis: English in Academic and Research Settings*. Cambridge: Cambridge University Press.

Swales, J.M. 1998. *Other Floors, Other Voices: A Textography of a Small University Building*. Mahwah, NJ: Lawrence Erlbaum.

Tarone, E., Dwyer, S., Gillette, S. and Icke, V. 1981. "On the use of the passive in two astrophysics journal papers." *The ESP Journal* 1: 123-140.

Villez, B. 1996. "Écrire comme un Anglais: Réflexions à l'usage des rédacteurs francophones dans le domaine juridique." *Anglais de Spécialité* 11-14: 447-453.

Vinay, J.-P. and Darbelnet, J. 1995. *Comparative Stylistics of French and English: A Methodology for Translation*. Philadelphia: John Benjamins.

Titles of English and German Research Papers in Medicine and Linguistics

INES-A. BUSCH-LAUER
University of Leipzig, LSP Centre

Introduction

Titles, abstracts and key-words are decisive instruments both for scientists and bibliographers to retrieve information and to establish reference to recently published research. Titles are first impressions of a research work and as T. Albutt (cf. Day 1989) has aptly remarked: "First impressions are strong impressions; a title ought therefore to be well studied, and to give, so far as its limits permit, a definite and concise indication of what is to come." Thus, it is the information content and the communicative effectiveness of titles that are clues for readers to decide whether a book, article or conference paper is worthwhile studying or not. Titles serve three basic functions: (1) They identify a piece of work (nominative function); (2) They designate the contents of a piece of work (designating function) and (3) They are appealing (advertising function).

But what criteria are required for a good title of scientific work? Style manuals for writing science often give inadequate answers to this question (cf. Day 1989, Ebel, Bliefert and Avenarius 1993, Staheli 1986). Day (1989: 15) emphasises the need for a short and concise title and defines it "as the fewest possible words that adequately describe the contents of the paper." Webster's Medical Secretaries Handbook (WMSH) (1979: 399) requires that the title should be "a clear, succinct, informative statement of the main topic of the article". Swales and Feak (1994: 205) suggest the following three requirements: (1) The title should indicate the topic of the study; (2) The title should indicate the scope of the study (i.e., neither overstating nor understating its significance); (3) The title should be self-explanatory to readers in the chosen area.

However, when we consider authentic titles we will easily recognise that they often do not correspond to these requirements. Day (1989: 16) clearly demonstrates that titles in medicine are often unspecified (cf. 1 and 2, no indication of subjects, disease, etc.), rather long and that they sometimes even suffer from syntactic incompatibility (cf. 3, double reference of resulting).

(1) **Action of** Antibiotics on Bacteria.
(2) **Action of Streptomycin on** Mycobacterium tuberculosis.
(3) Multiple **Infection** Among **Newborns Resulting from** Implantation with Staphylococcus aureus 502A (Day 1989: 17)

The same is true for linguistic titles which prove inadequate in terms of their semantic scope (cf. 4-5), their length (cf. 6-9) and with regard to the use of abbreviations (cf. 9).

(4) Action Research and Reading Difficulty (ESP J 11/2, 1996).
(5) Fächerübergreifende Sequenzen. Beispiele und Anregungen aus der Praxis (Praxis 43/4, 1996).
(6) Die Rolle fiktional-ästhetischer Medientexte im Englischunterricht: Medienkompetenz als Befähigung zu interkulturellem Verstehen (ZFF 7/2, 1996).
(7) Textmuster zur Erklärung hypothetischer Zusammenhänge in wissenschaftsjournalistischen Aufsätzen mit medizinisch-naturwissenschaftlicher Thematik (Deutsch -Englisch-Französisch) (GAL-Bulletin, 25, 1996).
(8) Unpeeling the Onion: Language Planning and Policy and the **ELT** Professional (TESOL Q, 30/3, 1996).
(9) Was ist Aufklärung - Für das **MfS**? Für die **SED**? Für alle? Gruppensprachlich und allgemeinsprachliche semantische Normen im Vergleich (GAL 1996).

In addition to this, one can observe a trend that titles of conference papers (often used for publication) do not convey any information about what the speaker is actually going to talk about and suffer from stylistic deficiencies (cf. 10).

(10) Die Janusgesichtigkeit von Übersetzungswissenschaft und funktionalem Übersetzen in der Sprachausbildung von Nichtphilologen in den Ingenieur- und Technikwissenschaften (GAL 1996).
(11) Der stumme Impuls: ein probates Mittel zur Förderung der Sprechfertigkeit (Englisch 1/1996).

While reading and checking the references to hundreds of articles and books for the *Kleine Bibliographie fachsprachlicher Untersuchungen* published in *Fachsprache - LSP Journal* it became obvious to me how difficult it is to classify scientific literature according to content areas by only relying on titles. Therefore, I started examining the communicative appropriateness of titles. The present interlingual study reports on an investigation into 150 German and English titles in two fields -linguistics and medicine- focusing on their length, structure and communicative effectiveness. The corpus material comprised 75 English titles (corpus 1) and 75 German titles (corpus 2).[1] Twenty-five titles each were taken both from medical and linguistic quality research paper journals and conference papers. These titles were designed by native speakers of English and German, respectively. To study L1/L2 differences a third corpus was compiled which consisted of 25 English titles written by German researchers. First, the three corpora were analysed with regard to the length of titles and their syntactic structure. In a second step, semantic aspects were included in the analysis.

Linguistic research on titles

To date linguistic research work on titles has focused on the various functions and types of titles (cf. Hellwig 1982, Hoek 1981, Sandig 1971); the relation between titles and text; the advertising function of titles (Rothe 1986); title reception from a psycholinguistic point of view (Schallert 1976, Schwarz and Flammer 1981) and the translation of titles (Nord 1993). However, LSP research on scientific titles is still rare (cf. Dietz 1995, Gnutzmann 1989 and Posteguillo 1997) though documentation and information of scientific research focus on their effectiveness to automatically retrieve and process information for data bases.

Christiane Nord (1993) was the first linguist to perform a comprehensive investigation into the pragmatic, intercultural and interlingual aspects of title translations in six genres (books, non-fiction, children's literature, poems, narratives, and title translations of scientific articles from four languages (German, English, French and Spanish). She (1993: 85pp.) exemplifies six basic functions of titles regarding the language functions described by Roman Jakobson. These are: (1) the distinctive function of titles; (2) the metatextual function; (3) the phatic function; (4) the nominating function; (5) the designating function, and finally (6) the appealing function of titles. Furthermore, Nord proposes to consider titles as a genre in its own right because titles can be viewed upon as very condensed metatexts covering all criteria which define textuality according to Beaugrande and Dressler (1981). Nord defines 'title' as follows (translation by Busch-Lauer):

> The title is a metacommunicative unit which is assigned the status of a text because it possesses, independently of the co-text, its own type of cohesion, coherence, intentionality, acceptability, informativity, and situationality. There is an interrelation (compatibility) between title and co-text which has to be established by the recipient. Therefore, the recipient has to acquire experience in

using titles which, in turn, is based on the conventions of these texts. (Nord 1993: 44p.)[2]

Claus Gnutzmann (1988) investigated titles from an interlingual and inter-subject point of view. He studied a corpus of 1335 English titles taken from 11 Anglo-American specialist journals of five disciplines: mathematics, economics, mechanical engineering, linguistics and education. To categorise titles occurring in the special languages he subdivides them into nominal, verbal and free structures. Nominal structures predominate in scientific writing and can be further subclassified (according to the degree of their modification) into unmodified, pre- and postmodified (or both), coordinated and free.

Concluding his study, Gnutzmann states (1988: 33) that the higher the degree of specialisation/abstractness of journals in a field the more condensed is the structure of titles. In addition, he demonstrated that there is a correlation between the degree of theoretization of a discipline (e.g. mathematics) and the structural density of titles.

In his doctoral thesis Gunter Dietz (1995) analysed the function, structure and semantics of titles in the German scientific communication studying material from four specialties: veterinary science, forestry, education and music. His corpus material (400 titles) was drawn from specialist journals, monographs and literary texts. In his analysis he investigated the length of titles, their syntactic structure, i.e. the proportion of titles/sub-titles; the proportion of nominal and verbal titles; the information content and rhetorical aspects of titles.

A very interesting frequency analysis of titles from computer science (and sub-disciplines), business and chemistry was presented by Santiago Posteguillo (1997). Focusing on the occurrence of prepositions, articles and non-finite verb forms, his study revealed that there are more statistically significant sub-disciplinary variations in the use of these structures in comparison with the disciplinary differences detected.

Results

Although a corpus study of 150 titles is by far too small for generalisations and conclusions, it may serve to describe the quality of titles in medicine and linguistics. The results also prove valid to derive recommendations for authors of conference papers and articles to improve their research titles. Finally, the study allows for a comparison between English and German titles in fields which are of particular relevance to German scientists who publish in English.

Length of titles

The length of a title is an indicator of the amount of information an author intends to give the reader prior to text reception. It is generally assumed that the longer the title the more information units there are to help readers decide on the relevance of the article/book for their own work. However, as we have seen from the above examples, a lengthy title may also confuse or disorient readers.

To measure the length of titles, all words which were included in a heading were counted. The concept 'word' was defined as the unit occurring between spaces. Abbreviations were treated as one word; compound words and hyphenated words were counted according to the number of their autosemantic components. Following these definitions the absolute number of words, the average rate of words per subject area and per individual corpus were determined as well as the maximum/ minimum number of words in each corpus. To be able to compare the data, the amount of titles having 1-10 words, 11-20 words and more than 20 words were scored (absolute number and in per cent). A detailed picture of these results is given in table 1.

The investigation has revealed that titles in medicine contained an average of 9.9 words per title, whereas linguistic titles only contained an average of 8.4 words. In the study of Dietz (1995) titles from veterinary science were the longest (11 words per title; literary titles were the shortest (5 words per title, cf. Nord 1993: 106). The differences between the length of titles in medicine and linguistics become more apparent when we consider the average number of words in the individual corpora - the number of words in all three linguistic corpora is smaller than in the medical corpora. This difference may be due to the fact that articles in the natural sciences and medicine require more detailed information, e.g. on the type of investigation, the type of disease, the patients, research methodology. Linguistic titles, on the other hand, do not follow such strict subject conventions.

German titles in both the medical and linguistic corpus are shorter than the English L1 and English L2 titles. This relative shortness is due to the fact that authors often use compound words.

Table 1. Length of Titles (in Words) in the Medical and Linguistic Corpora

Subject	MEDICINE			LINGUISTICS		
Corpus Details	English Corpus (25 titles)	German Corpus (25 titles)	L2 Corpus (25 titles)	English Corpus (25 titles)	German Corpus (25 titles)	L2 Corpus (25 titles)
- absolute # of words	280	210	255	227	204	214
- maximum # of words	25	19	18	16	13	21
- minimum # of words	2	3	4	4	4	3
- average # of words per subject corpus	9.9 (total medical corpus)			8.4 (total linguistic corpus)		
- average # of words per individual corpus	11.2	8.4	10.2	9.08	8.1	8.5
-range of words/title						
1 - 10 words	17=68%	18=72%	14=56%	19=76%	19=76%	20=80%
11 - 20 words	4= 16%	7 =28%	11=44%	6=24%	6 =24%	4 =26%
> 20 words	4= 16%	0	0	0	0	1=4%

The criterion 'range of words' was introduced to determine the amount of titles the length of which coincides with the recommendations of style manuals (8-10 words maximum per title). As can be seen from table 1, the majority of the titles conform to the proposed norm: an average of 65% of the medical titles (having up to 10 words); an average of 77% of the linguistic titles. However, when we consider the individual corpora we discover differences, e.g. 56% of the L2-titles in medicine use up to 10 words, whereas about 80% of the linguistic L2-titles range in the category 1-10 words. Interestingly, authors of titles in linguistics almost never run over a limit of 20 words per title, which would actually curb readability.

The syntactic structure of titles

According to Dietz (1995) and Rothe (1986: 17p.) we can subdivide titles into titles proper, i.e. monostructured *(eingliedrige)* titles and title and sub-title, i.e. bistructured *(zweigliedrige)* titles. The study of Dietz, for example, revealed that 90% of the articles in veterinary science and 80% of the articles in forestry only consist of a monostructured title. In contrast, about 50% of the articles in education consisted of both a title and a sub-title and so did about 2/3 of the articles in music.

A second criterion to classify titles syntactically is the structure of the title components. Dietz (1995: 21pp.) distinguishes between nominal titles[3], i.e. titles which comprise at least 1 nomen which is superordinate (head) to the other structures in the title and verbal titles (Dietz 1995: 22; cf. Gnutzmann 1988),[4] i.e. titles that contain a verb as the predominating element in a title, as well as prepositional titles and adjectival/adverbial titles.

Table 2 summarises the results for the studied medical and linguistic corpora with regard to their structure (amount of monostructured titles and titles-subtitles) and the classification into nominal and verbal titles.

Table 2. Structure of Titles (for 25 Titles in Each Corpus)

Subject	MEDICINE			LINGUISTICS		
Type of Structure	English Corpus	German Corpus	English L2 Corpus	English Corpus	German Corpus	English L2 Corpus
Monostructure	19 = 76%	9 = 36%	14 = 56%	9 = 36%	11 = 44%	18 = 72%
- nominal	18	9	14	7	11	12
- verbal	1			2		6
Title/Subtitle	6 = 24%	14 = 64%	11 = 44%	16 = 64%	14 = 56%	7 = 28%
- nominal	5	14	11	11	14	6
- verbal	1			5		1

As we can see from table 2, authors in medicine prefer monostructured titles in English (both in L1 and L2). However, German writers tend to use more title-subtitle structures in German to convey the basic information of their articles both in medicine and linguistics. A trend towards this construction can also be assumed for the entire linguistic corpus. Both the English L1 and the German corpus preferred the title-subtitle structure, whereas authors of the English L2 titles predominantly used monostructured titles. This difference may be explained by the fact that most of these titles resulted from papers presented at conferences where authors had to follow the instructions of the conference conveners to submit short titles and thus, they kept the titles for publication.

A third criterion to be considered in the syntactic analysis is the relationship between title and subtitle from a structural and from a semantic point of view. Table 3 briefly summarises the results for the title (T) - subtitle (ST) relations of which four occurred in the corpus: (1) T + ST = nominal; (2) T = nominal + ST= verbal; (3) T = verbal + ST = nominal; (4) T = verbal + ST = verbal.

Not surprisingly, most titles have a nominal structure in both title and subtitle. In the medical corpus almost all titles have a nominal structure in all three corpora. The subject and contents in this field require unemotional informative titles which are best expressed by nominal structures. Therefore, verbal constructions and questions remain rather an exception than a rule (cf. 12). Linguistic articles, on the other hand, seem to be more open to accept verbal titles and questions, e.g. in the English L1 corpus, which points to the creativity of authors. Still, these structures are only used to point to the argumentative character of the topic under discussion (cf. 13).

(12) Ulcusheilung durch Heliobacter-pylori-Eradikation: Genügt eine Woche Therapie? (GMC-18).

(13) Collaborating with Content-Area Teachers: What We Need to Share (ELC-18).

These results correlate with those revealed by Dietz (1995) for the natural and social sciences. The majority of the studied monostructured and bistructured titles in his corpus have a predominantly nominal structure (cf. table 4 quoted from Dietz 1995: 24). In contrast, 16% of the literary titles in his corpus follow a verbal structural pattern.

Table 3. Relation between Title (T) and Subtitle (ST)

	MEDICINE			LINGUISTICS		
	English Corpus (6 titles)	German Corpus (14 titles)	English L2Corpus (11 titles)	English Corpus (16 titles)	German Corpus (14 titles)	English L2Corpus (7 titles)
T nominal ST nominal	5	12	11	8	12	5
T nominal ST verbal	-	2	-	4	2	1
T verbal ST nominal	1	-	-	3	-	1
T verbal ST verbal	-	-	-	1	-	-

Table 4. Syntactic Structure of Titles - Proportion of Nominal/Verbal and Other Structures in Titles and Subtitles (Source: Dietz 1995: 24; Data for Articles)

Group of article (#)	Main Title			Subtitle		
	Nominal %	Verbal %	Others %	Nominal %	Verbal %	Others %
VZ (18)	100.0	0.0	0.0	100.0	0.0	0.0
FZ (35)	91.4	5.7	2.9	97.1	2.9	0.0
PZ (85)	85.5	10.7	3.5	96.4	3.6	0.0
MZ (136)	77.9	14.7	7.4	95.6	2.2	2.2
L (149)	75.2	16.1	8.7	99.3	0.0	0.7

VZ = veterinary articles; FZ = articles from forestry; PZ = education; MZ = articles from music science; L = literature, # = number of studied titles

Another interesting criterion which describes title-subtitle relationship is their semantic correlation. Dietz (1995: 134pp) used five categories of main title-subtitle relations in his study of German scientific titles. These are: (1) General title - specific subtitle; (2) Title as subject - subtitle as predicament; (3) Subtitles with anaphorous reference to title; (4) Thematic title- non-thematic subtitle; (5) Vague/ Miraculous title - informative subtitle.

In the studied medical and linguistic corpora there is no marked linguistic reference expressed between title and subtitle. Subtitles are separated from titles by colon (English titles) or hyphen (predominating in German texts). In addition, the importance of main titles is underlined by lay-out, size and type of print. Thus, large letters catch the attention of readers; the smaller print of subtitles specifies the investigation.

In both the medical and linguistic corpora I could observe the trend that (main) titles convey more general information, i.e. in most cases they are superordinate, and subtitles add information or specify general notions, i.e. they are subordinate.

In the medical corpus, the following generalised semantic components of titles and subtitles could be observed in both the English and German languages:

Table 5. Components of Title-Subtitle Structure in Medicine

General Title	Specific Subtitle: Specification regarding
- Type of disease/symptom studied - Groups of patients studied, - Type of operation/procedure - Relation between diseases/symptoms	- Research methodology - Type of study - Type of text - Aspects/ implications for therapy, cure

(14) Apolipoproteins and Ischaemic Heart Disease: Implications for Screening (EMC-9).

(15) -2-Makroglobulin im Urin. Differentialdiagnostische Bedeutung bei Rejektionen und Infektionen nach Nierentransplantation (GMC-14).

(16) Bleeding Peptic Ulcer: An Audit of Conservative Management (E2MC-1).

In the linguistic corpus one can trace two main trends (cf. tables 6.1 and 6.2). One stream of titles proceeds from general to specific information but remains rather vague, the other proceeds from a sample or a fuzzy title to a more informative subtitle. Sometimes these two trends also overlap. However, in contrast to the medical corpus some titles in linguistics would not be comprehensible without their subtitle.

Table 6.1. Components of Title-Subtitle Structure in Linguistics

Title: General aspects/topic	Subtitle: Specification regarding
- Theory/practice - Type(s)/aspects of language studied - Process	- Aim of study - Process (e.g. teaching) - Type of consideration/text

(17) Metacognitive Strategies in Second Language Academic Reading: A Qualitative Investigation (ELC-12).

(18) Zum Verstehen von Lehrbuchtexten - Eine Untersuchung mit Auszubildenden im Berufsfeld Körperpflege (GLC-10).

(19) LSP in Diachrony: A Comparison of two German-Sorbian Dictionaries (E2LC-25).

Table 6.2. Components of Title-Subtitle Structure in Linguistics

Title: Nebulous/Pretentious Phrase   **Subtitle**: (Precise) Information on- - Quotation   - Type of study, process or topic - Example or Incomprehensible aspect

(20) Their Name Are Juan and Rosa: Understanding and Responding to Noun Number Errors in ESL Writing (ELC-20).

(21) Prüfungstexte auf dem Prüfstand: Ein interdisziplinärer Ansatz zur Untersuchung der Fachlichkeit (GLC-13).

(22) Familiar Problems, Unorthodox Solutions: How Lexicography Can Aid LSP Application (E2LC).

As the semantic components recur in all three subcorpora of the disciplines, one can assume that the formulation of titles largely depends on the subject to be discussed and the conventions of a discipline. More generally speaking, there are differences between the information content of titles in the natural sciences and the social sciences that are basically due to the subject fields and the research methodology applied in them (experimental approach in medicine vs empirical approach in linguistics).

Information content of titles

Titles of scientific research function as a condensed frame of the article proper. The information content should be sufficient to enable readers to get the main idea of the text. Using examples from both subject corpora, I shall first examine the informativeness of titles and then compare titles of both subject areas. Within this context I will also discuss some linguistic and interlingual aspects of titles.

Titles (23-28) are taken one from each medical and linguistic corpus and seem to be prototypical for each discipline. Although the chosen medical examples are quite long and their readability is restricted, they convey very precise information to the reader. The information content is dense and high. It is clearly defined in the titles what the article text will be about; (disease, e.g. (23) *pancreatitis*; patients, e.g. (23) *Phase I AIDS- patients*; methods applied, e.g. (23) *treatment with Didanosine*). Still, we can find redundant words in some of the medical titles, e.g. *analysis of, associated with the development of*, which could be deleted so that title (23) for example reads as follows: *Potential Risk Factors for Pancreatitis in Phase I Patients with AIDS or AIDS-related Complex Receiving Didanosine* (17 instead of 23 words).

To dissolve information density, a subdivision into title-subtitle structures would be indicated in cases where the number of words in a monostructured title runs over 15-20 words. Title (25) follows the title-subtitle pattern and is a well-structured example. The title presents the key topic, e.g. *bleeding peptic ulcer*, the subtitle conveys the type of investigation, e.g. *audit, conservative management*. In contrast to the English L1 corpus, this method is commonly applied by German researchers in both the German and English language.

(23)   Analysis of Potential Risk Factors Associated with the Development of Pancreatitis in Phase I Patients with AIDS or AIDS-Related Complex Receiving Didanosine (EMC-15).

(24)   Antibiotische Prophylaxe bei endoskopischen Eingriffen an den Gallenwegen (GMC-21).

(25)   Bleeding Peptic Ulcer: An Audit of Conservative Management (E2MC-1).

Linguistic titles usually do not convey these informative aspects, often they keep the topic rather vague and unspecified (cf. 26-28). This approach may be valuable for literary titles because here authors can rely on readers' curiosity to explore a text for pleasure. However, it does not help the busy teacher who is basically addressed by linguistic articles of this type. Title (26) *If You Can Read This, Thank TV* opens up speculations on what the article will be about. Is it reading comprehension, or TV education, is there any relation between reading and TV? Although the title is short and the readability appropriate, the message conveyed remains unclear. It seems as if the reader is forced to have a closer look at the text to 'dig out' what the author safely hides in the title. The same is true for title (27). These types of titles either force readers to develop phantasy or distract them completely from reading. As we have already seen, quite a number of these titles appeared in the German corpus and they do not fulfil their communicative function. Although linguistic debate is about comprehensibility of texts, linguists themselves often do not consider this aspect when writing their texts.

Title (28) is a stylistically inappropriate and wordy title. *LSP research* seems to be the key-element. What the author wants to write about is the development of LSP research in the early 1990s. Thus, an efficient version of the title would be *LSP Research in the early 1990s*. This revision deletes 8 words and it also provokes readers' interest.

(26)   If You Can Read This, Thank TV (ELC-19).

(27)   Pädagogisierung des Fremdsprachenunterrichts. Schritte in Richtung zeitgemäßen Lernens (GLC-21).

(28)   The Interdisciplinary Stage of Development of LSP Research at the Beginning of the 90ies (E2LC-14).

Thus, we can conclude that linguistic titles are often not precise enough, they 'hide' the contents of a text, they do not indicate the type of text and kind of research performed.

In medicine it is no longer necessary to classify texts in the title because journals usually publish material according to genres (original article, review article, case report). Nonetheless, readers of medical titles can often derive the text type already from the title (cf. 29-31). They may decide prior to reading whether the described approaches touch their own subject area of interest or not.

(29) Thrombolysetherapie des akuten Herzinfarktes - derzeitiger Stand und neue Entwicklungen (*review article*) (GMC-4):

(30) Intravascular Lymphomatosis: A Clinicopathological Study of Three Cases (*case study*) (E2MC-7).

(31) Symptomatik und endokrinologische Befunde bei katecholaminsezernierenden Tumoren. Ergebnisse bei 106 konsekutiven Patienten (*experimental study*) (GMC-11).

Medical titles refer to performed studies and their results rather than to a description of projects, etc. (as in linguistics). Therefore, we can find lexemes describing the notion of 'result' and 'consequence' in the medical corpora. In addition, author(s) of medical titles often signal the purpose of their study by indicator words (words for diseases; types of investigation), e.g.

> **English corpus**: *commentary on ...; incidence of ...; progression of ...; implications for screening/cure/ therapy, the influence of (agent)... on (patients, disease) ...; prevention of ...; therapy in ...; a case of ...; critical review of ...; clinicopathological considerations, impact of...;*
>
> **German corpus**: *Therapie des ... bei; Diagnosesicherung durch ...; Symptomatik und Befunde bei ...; (Operationsmethode) bei (Erkrankung) ...; Rejektionen, Infektionen, Prophylaxe bei ...; Unverträglicheit bei ...; Primärmanifestation von ...;*

In contrast, linguistic titles include 'experience' and 'empirical thoughts' and denote the 'process of research' rather than results. In English this process character is often expressed by the gerund, e.g. *designing courseware for* (...). In German authors use nominal structures (e.g. *Schritte in Richtung* ...) and prepositions (e.g. *von ... zur*) to indicate the notions of development and progress, e.g. **Von der Terminographie zur Textographie. Computergestützte Verwaltung textsortenspezifischer Versatzstücke.** The preposition *zu/zur* is also used as a 'hedge' to indicate that the article will restrict itself to selected aspects of an issue. The English equivalent *on* did not occur in the whole corpus except for two linguistic English L2-titles: a conference paper title (*On the Use of Tense and Aspect...*) which was transferred for publication into a gerund structure (*Using Aspect ...*); the second title was due to translation (*On the Discourse Function of Full Inversion in English / Zur Funktion der ...*).

A common type to form monostructured titles in both disciplines and across languages is to express the probable interrelation of two or more key-concepts using the connector *and/und*. In medicine this type of titles indicates review articles, in linguistics it indicates argumentative essays, cf. (32-34) versus (35-37).

(32) Cholesterol **and** Violent Behaviour (EMC-21).
(33) Cold Urticaria, Raised IgE **and** HIV Infection (EMC-25).
(34) Fettarme Diät **und** körperliches Training bei koronarer Herzkrankheit (GMC-12).
(35) Pragmatist Discourse **and** English for Academic Purposes (ELC-8).
(36) Fachspezifisches Formulieren **und** Rezipieren (GLC-25).
(37) Discourse Markers **and** Conversational Coherence (E2LC-15).

As we have seen from the analysis, the differences described in the information content of titles between medicine and linguistics are basically due to the conventions of the discipline, i.e. the subject of investigation, the applied research methods, the type of text. Although the medical titles are relatively long, they convey the necessary information for automatic information retrieval and data bases, like *Current Contents, Index Medicus* because they are precise and their structure is clearly defined by the contents of a text. No matter whether they are monostructured or consist of title and subtitle, they serve the communicative function of a metatext. The readers are well-informed from the title which is often separated from the article text or abstract. Linguistic titles, on the other hand, did not always prove effective enough with regard to their information on the following text. These insufficiencies are related to the structure of the title, to the vague contents and wordiness of titles.

Stylistic and rhetorical aspects

Titles of research designate contents and they also serve as eye-catchers. A brief examination of rhetorical devices used in the titles of the two subject fields underlines the impersonal and unemotional style of medicine and a more creative one in linguistics. It is the topic of an article itself that provokes the interest of medical researchers. It is the linguistic peculiarity that catches linguists as readers. In linguistics is seems to be a matter of creativity and research prestige to use outstanding, unusual titles. Thus, we can assume that there is a different reading expectation which has traditionally been formed by the convention of a given discipline. In contrast to the medical corpus, the linguistic titles contained some rhetorical devices which were either introduced by chance or on purpose, like elements of prosody, repetition, antonymy and even intertextuality, e.g.

(38) Business is Booming: Business English in the 1990s (ELC-5).
(39) Prüfungstexte auf dem Prüfstand. Ein interdisziplinärer Ansatz zur Untersuchung der Fachlichkeit (GLC-13).
(40) Modal Meanings in the Structure of Argumentation (ELC-23).

However, we can also find stylistic deficiencies that hamper readability and should be avoided:

1. The studied English L1-titles, for example, contained a number of abbreviations that might not be comprehensible to the general reader, e.g. CALL, ESP, ESL.
2. Some of the German titles seemed to be stylistically deficient because of their complex syntactic structure, e.g. *Zum Begriff des Technical Writing als Intertextualität schaffendem Prozeß.*
3. The use of foreign words and uncommon words also decreases readability of a title, e.g. *Fremdsprachenforschung als Vexierbild.*

As all these examples were taken from published articles of conference papers, we may conclude that once the title of a research paper has been fixed, linguists are unwilling to find a better title for publication. In contrast to multi-author articles in medicine, it is interesting to note that titles in linguistics still remain an expression of individual style. In some cases you can even decipher the author of an article because of a certain unusual style in formulating titles. It would be interesting to find out whether these unusual individual styles also contribute to principles of title formulation in a given discipline or whether restrictions of journals predominate this process.

Interlingual aspects

Finally, some remarks on the effectiveness of titles in the L2-corpora. In medicine German authors are forced to submit English titles and abstracts of their research work. Thus, most authors have to rely on their ability to translate these titles, which may result in less effective L2-titles. It is therefore advisable to consider the different title structures in both languages and to search for a functionally appropriate translation (Nord 1993).

(41) Pathogene und apathogene Immundefizienz-Viren: Pathogenitätsdeterminierende Faktoren (**Translation:** Factors Determining the Pathogenicity of Immunodeficiency Viruses)

The English translation of (41) conveys the information content even better than the original because of transforming the title-subtitle structure into a monostructure which focuses the attention of readers on *pathogenicity*.

The L2-titles in the linguistic corpus reflect the difficulty to translate rather vague and abstract notions. Thus, some of the titles present with lexical and stylistic inadequacies, e.g.:

(42) **Abstracting in the Perspective** of Producing a Text (2x -ing-form, **in** the Perspective).
(43) **Peculiarities** of News Agency **Copies** (in Contrast to Newspaper Texts) (German: Besonderheiten von ... im Kontrast, im Unterschied zu...)
(44) **Evaluating Acts** in **Economic** Texts **from** German Newspapers (Bewertungshandlungen in Wirtschaftstexten aus ...)

Summary

The investigation has clearly shown that titles in both disciplines are first impressions of a piece of work, they are the 'entrance' to a document. This 'entrance' differs across disciplines and may also vary within the field under investigation. This is mainly due to the disciplinary conventions, sometimes set by the publication authority, to the communicative framework of the document (purpose, addressee, competition among scientists, journals, etc.), the contents of the document as well as the author's personal writing/ reading experience which are also influenced by his/ her cultural background.

By nature of the discipline, titles in medicine are long, precise and informative. Medical researchers and doctors have to cope with a flood of new publications, constantly being under time pressure, so the title of a document is of utmost relevance for them, both for scanning and selecting research work. The more precise the title is, the easier it is also for bibliographers to compile data for indexing, abstracting and other documentation purposes.

Titles in linguistics, on the other hand, are rather short and often remain vague and abstract. In contrast to the collective authorship in medicine, linguists tend to be single authors, which allows them to express their individual style, e.g. using metaphors and questions. Often linguistic articles result from conference papers and thus authors keep the titles, although these have been rather vague and 'preliminary'. As review processes and competition for publication are not as strict as in medicine, publishers and readers usually do not mind linguistic titles in this form. However, more precise titles would be helpful for both linguists and bibliographers to trace recent developments. Another problem, the analysis pointed to, is that German researchers (and perhaps other non-native speakers of English) use conventions of L1 to formulate titles in English, which are often inappropriate in terms of language and style. Therefore, scientists of both disciplines should be aware of the fact that both titles in L1 and L2 serve communicative functions. They should be informative condensed 'appetisers' to a text.

Notes

1 The material was taken from: English for Specific Purposes (ESP J); TESOL Journal; Fachsprache (FS); Zeitschrift für Fremdsprachenforschung (ZFF); GAL-Bulletin; British Medical Journal (BMJ), Journal of the Royal Society of Medicine, The Lancet, The Journal of Infectious Diseases, The New England Journal of Medicine, Cancer, The American Journal of Medicine, Cardiology, Deutsche Medizinische Wochenschrift, Internist, Chirurg, Innere Medizin, Therapiewoche, various conference abstracts in these fields.

2 German: "Der Titel ist eine metakommunikative Einheit, der wir Textstatus zusprechen, weil sie unabhängig vom Ko-Text eine je eigene Form von Kohäsion, Kohärenz, Intentionalität, Akzeptabilität, Informativität und Situationalität aufweist. Zwischen Titel und Ko-Text besteht ein Zusammenhang (Kongruenz), der vom Empfänger hergestellt werden muß. Dafür braucht der Empfänger eine Erfahrung mit Titeln, die wiederum auf der Konventionalität dieser Texte beruht." (Nord 1993: 44f.).

3 Dietz (1995: 21) presents the following as an example for a nominal title: "Klimaänderung durch erhöhte Spurengehaltsstoffe in der Atmosphäre."

4 Dietz (1995: 22) presents „Soll der Mensch Schöpfer spielen?" as an example of a verbal title.

References

Beaugrande, R. de and Dressler, W. 1981. *Einführung in die Textlinguistik*. Tübingen: Niemeyer.

Day, R. A. 1989. *How to Write and Publish a Scientific Paper*. 3rd edition. Cambridge: Cambridge University Press.

Dietz, G. 1995. *Titel wissenschaftlicher Texte*. (Forum für Fachsprachenforschung 26). Tübingen: Gunter Narr.

Ebel, H. F., Bliefert, C. and Avenarius, H. J. 1993. *Schreiben und Publizieren in der Medizin*. Weinheim, New York, Basel, Cambridge, Tokyo: VCH.

Gnutzmann, C. 1988. "Aufsatztitel in englischsprachigen Fachzeitschriften. Linguistische Strukturen und kommunikative Funktionen." In C. Gnutzmann (ed.), *Fachbezogener Fremdsprachenunterricht*. (Forum für Fachsprachenforschung 6). Tübingen: Gunter Narr, 23-38.

Hellwig, P. 1982. "Titulus oder zum Zusammenhang von Titeln und Texten. Titel sind ein Schlüssel zur Textlinguistik." In K. Detering (ed.), *Sprache erkennen und erleben.* Tübingen: Niemeyer, 157-167.

Hoek, L. H. 1981. *La marque du titre. Dispositifs sémiotiques d'une pratique textuelle.* (Approaches to Semiotics 60). La Haye, Paris, New York: Mouton.

Nord, Ch. 1993. *Einführung in das funktionale Übersetzen. Am Beispiel von Titeln und Überschriften.* Tübingen, Basel: Francke.

Posteguillo, S. 1997. *Writing Titles for Computer Science Research Articles: Sub-Disciplinary Variations in Academic English.* (Paper delivered at the 11th European Symposium on Languages for Specific Purposes, 18-22 Aug, 1997, Copenhagen, in print).

Rothe, A. 1986. *Der literarische Titel. Funktionen, Formen, Geschichte.* Frankfurt/M.: Klostermann.

Sandig, B. 1971. *Syntaktische Typologie der Schlagzeile. Möglichkeiten und Grenzen der Sprachökonomie im Zeitungsdeutsch.* (Linguistische Reihe 6). München: Hueber.

Schallert, D. L. 1976. "Improving Memory for Prose: The Relationship between Depth of Processing and Context." *Journal of Verbal Behavior and Verbal Learning* 15, 621-632.

Schwarz, M. N. K. and Flammer, A. 1981. "Text Structure and Title - Effects on Comprehension and Recall." *Journal of Verbal Behavior and Verbal Learning* 20, 61-66.

Staheli, L. T. 1986. *Speaking and Writing for the Physician.* New York: Raven Press.

Swales, J. and Feak, C. B. 1994. *Academic Writing for Graduate Students. A Course for Nonnative Speakers of English.* Ann Arbor: The University of Michigan Press.

Webster's Medical Secretaries Handbook 1979. A. H. Soukhanov et al. (editorial board). Springfield, Massachusetts: Merriam-Webster Inc.

Genres and the Media

That's not News:
Persuasive and Expository Genres in the Press

TORBEN VESTERGAARD
Dept of Languages and Intercultural Studies
Aalborg University

Introduction

The prototypical newspaper text is the news report. This is by far the predominant genre in the training of journalists, and it is certainly also the text type that has attracted the bulk of scholarly attention. Thus one major book on newspaper language from the early 1990s, Bell (1991), after having established the generic distinction between three kinds of newspaper copy, viz. service information, opinion, and news, is exclusively concerned with the latter. Another major book from the same year, Fowler (1991), does contain a chapter on leading articles (ch. 11), but symptomatically, the chapter does not give a single reference. What the news report (purportedly) does is simply to give an account of events as they actually happened, and if we were required to give a characterization of the newspapers' task in one sentence, most of us would probably agree that this is what newspapers should do; but in addition, any newspaper will contain texts that cannot possibly be understood as trying to relate news about events, but which, rather, carry comments, interpretations, evaluations, recommendations, etc. based on and relating to the events and issues reported in the news texts proper. It is with these texts that are expressly *not* news that this paper is concerned.

As is well known, the separation of "report" and "comment" holds a central place in the ethos of the journalistic profession, the crucial argument being that readers should be able to tell when the newspaper is giving an "impartial [...] picture of what was really happening" (Andrew Neil in Wilsher, Macintyre and Jones 1985: x), and when they are expressing points of view. Discourse analysts of the critical school have long claimed that the profession's allegiance to the distinction is ideological (cf. Fowler et al. 1979, Fowler 1991, van Dijk

1988, White 1997), and others have pointed out that there is no such thing as a neutral and impartial account of "what really happened", since all representation ineluctably presupposes interpretation and selection (Stubbs 1996). Nevertheless, the distinction is strictly adhered to, at least in quality papers (see further below), not just in that the non-news sections of the newspaper are clearly marked off as being different from the news sections proper, by typographic and other means, but also, and more interestingly, in some obvious linguistic choices, even at clause level (Vestergaard 1996), cf. below.

In this paper I shall look at these non-news newspaper texts from a genre theoretical point of view: I shall discuss in what ways they can be said to form a genre of their own, distinct from the prototypical newspaper text, and in so doing I hope to have something to say on the definition of "genre". In particular, I shall argue that, in spite of much current practice (cf. eg. Stubbs 1996), there may well be a need for a consistent terminological and conceptual distinction between "text type" and "genre".

Texts and genres

As genres are categories of texts with perceived shared characteristics, it may be in order to open this discussion of newspaper genres with a brief look at the concept of text. There are, to my knowledge, two ways of defining "text", an objectivist and a functionalist. According to the *objectivist* view, a text is simply a string of (ostensibly) linguistic symbols fixed on paper (or some other medium) "capable of being employed in the construction of meaning" (Bex 1996: 78). The corollary of this view is that there may be meaningless texts, a consequence that Bex himself is quite prepared to accept (op.cit. p. 79), and that a person can recognise a given object as a text even if he knows neither the language, nor indeed the notation system employed (ibid. p. 77). This is probably also the everyday sense of the word: we talk about Etruscan texts although no-one is able to decipher them completely.

According to the *functionalist* or semanticist view, on the other hand, a text is not just any string of linguistic symbols, but a sequence with a recognizable communicative purpose (which, of course, is not to be confused with the composer's communicative intention). This view, which is by far the most widely accepted among linguists (Halliday and Hasan 1976, 1985/89, Martin 1992, Mann, Matthiesen and Thompson 1992, Longacre 1992), has a number of corollaries. One is that certain ostensible texts on closer inspection may very well turn out to be "a collection of unrelated sentences" (Halliday and Hasan 1976: 1). Another is that to qualify as a text, a linguistic sequence should be reducible to one (macro-) proposition (Thompson, Matthiesen and Mann 1992, Longacre 1992). Finally, what to some readers may be an utterly incomprehensible collection of words and sentences may be a perfectly sensible and coherent text to others. These others are the privileged readers for whom the text is produced, its "imagined readers" (Coulthard 1994) or its "discourse community", as Swales (1990) puts it. From this it follows that, within the functionalist view on text, it is only with reference to a specific discourse community that the question of the textuality of a given piece of writing can be decided upon.

We are all of us members of several discourse communities, as Bex (1996: 169) points out, and we are thus all of us qualified readers of diverse texts. How we respond to them is, as our response to all other phenomena we come across, an effect of a) the properties of the phenomenon itself and b) our expectations based on previous experience. Now, texts are communicative events (Swales 1990: 58). And as, on the basis of previous experience, we classify events with certain shared salient characteristics into event types, also known as frames or scripts (cf. Tannen 1979), and as our response to new events is partly conditioned by the script activated by the event in question, so we group texts with perceived shared communicative purposes into text types, also known as genres. Our reaction to any given text, now, is in large measure conditioned by our previous experience of the genre to which we assign it. As a familiar example of this, consider how our reactions to advertisements posing as editorial material or as strip cartoons are clearly conditioned by our knowledge that in spite of certain outward appearances, the texts in question are in fact advertisements (cf. Vestergaard and Schrøder 1985: 62-65).

If recognisable communicative purpose is the defining characteristic of text, a reasonable starting point for generic classifications would be exactly communicative purpose as distinct from surface linguistic features, cf. Bhatia (1993: 86-88). A central attempt in that direction, and in addition an attempt to base a generic classification on deductive rather than on inductive reasoning, is to be found in Longacre (1972: 200ff.). Longacre distinguishes four main genres on the basis of two pairs of concepts: +/- temporal succession and +/-projection (roughly corresponding to future vs. non-future), thus:

Table 1

	- projection	+ projection
+ temporal succession	*narrative*	*procedural*
- temporal succession	*expository*	*hortatory*

Narrative texts recount a sequence of events represented as having taken place in the past (note how even science fiction texts use the past tense about the "narrative present"); procedural texts prescribe sequences of actions which must be followed to make an appliance operate sucessfully or to make a collection of ingredients unite into a dish of food; expository texts describe existing states of affairs and/or problems and their possible solutions, and finally, hortatory texts induce the reader to take some future course of action or to adopt an attitude or point of view. In (1992) Longacre adds one further macro-genre to his taxonomy, viz. *persuasive*, which shares features with both expository and hortatory texts: like the former it describes a problem and proposes solutions, but like the latter it also attempts to prevail upon the reader to accept the proposed solutions.

Any genre, in Longacre's view, is characterized by a particular configuration of "deep structure Moves". These Moves are clearly of a different conceptual status than the units referred to as "Stages" by Eggins and Martin (1997), in that the Stages, as the name implies, correspond to actual textual segments. Similarly Longacre's moves seem to operate on a far more abstract level than the units referred to by the same term in Bhatia (1993). Thus, they do not enter into any simple one-to-one correspondence with actual textual segments (cf. also Hoey 1994), and moreover, Longacre's model of genre, like that of Coulthard (1994) and Hoey (1994), allows for both embedding and recursion.[1] For instance, the "presentation-of-problem" move of a hortatory text may well be realized by a sequence of narrative text (in fact, this is not at all infrequent in advertisements, cf. Vestergaard and Schrøder 1985: 50-54).

Leaving aside procedural texts, which are irrelevant for present purposes, we are left with the following four macro text types, each of which, in its full form, is characterized by four "deep structure moves" as indicated:

<u>Narrative</u>
incident
tension
climax
denouement

<u>Expository</u>
problem
proposed solution
supporting argumentation
evaluation of proposed solution

<u>Persuasive</u>
problem

proposed solution
supporting argumentation
appeal to give credence/adopt values

Hortatory
establish authority
present problem
issue commands (possibly mitigated)
create motivation

I shall leave the question open as to the theoretically motivated number of possible macro text types and merely note that for the purposes of the present exposition we shall need yet another type, *descriptive*.

On the face of it this view on genre and text has some highly counter-intuitive, but theoretically extremely interesting, consequences, in that it cuts right across the concept of register and will classify texts from differing registers as belonging in the same genre, and texts from the same register as belonging in different genres. Thus, the cooking recipe would belong in the same text type as the operating instructions for a heavy chain saw, and the political news report would belong in a different text type from the political leader/editorial but in the same text type as the novel or oral narrative. Following Martin (1992: 546-573) I shall say that generic distinctions occur at the intersection of communicative purpose and register, which will at least land the recipe and the operating manual in different genres. Anyway, it is of course only from a narrowly register oriented perspective that the view that the news report is more closely related to story-telling than to other news-genres is surprising. From other points of view, the narrative nature of news reporting can be regarded as firmly established, cf. Bell 1991, Bennett 1982, Hartley 1982, Katz 1987.

One of the things to be considered in what follows is to what extent a generic subclassification of newspaper texts can be based upon register variables.

The leader page

In the four British national upmarket newspapers (i.e. the *Daily Telegraph*, the *Guardian*, the *Independent*, the *Times*), the material in the focus of interest of the present article occurs in canonical form on a double page near the middle or end of the (first section of the) newspaper. The contents and lay-out of these two pages are strikingly similar in all four papers: they contain, printed under the paper's logo, its leading articles on two or three topics and on the same page letters to the editor (in the *Guardian* on the left-hand page, in the three others on the right-hand page). In addition there is a large cartoon, and in all four except the *Independent*, comments by named columnists plus a gossip column containing a mixture of news and comments on trivial topics, narrated in a tongue-in-cheek fashion (the *Independent* carries these items in the two following pages). These pages, then, are the place where the newspaper and its readers talk to each other about serious subjects as well as trivia, the way educated middle-class people will.

The pattern of the upmarket papers recurs in the midmarket, *Daily Mail* and *Daily Express,* with the leading article facing a comment bearing the columnist's name and photo. Variations on this combination of leader(s), comments and gossip column also recurs in various forms in several of the populars (e.g. the *Mirror*, the *News of the World,* the *Sun*), whereas in the *People,* the leader is almost indistinguishable from the rest of the editorial material and is often placed directly adjoining the news report it relates to. In spite of the similarities, however, there are also several features differentiating the upmarket papers from the mid- and downmarket ones: first, in the populars, the leader page occurs early on in the paper (in the *Sun*, directly facing the "page three girl", who is actually to be found on page seven). Second, these papers have no qualms about printing trivia, and the ironic tone characterizing the gossip columns of the upmarket papers is absent. Third, the juxtaposition of leader and letters to the editor is absent (except in the *News of the World*), and the atmosphere of dialogue among equals aimed at in the leader pages of the upmarket papers is thus absent from the populars. As, in general, the distinction between news and comment is much less clear-cut in the populars than in the qualities (*pace* Hodge 1979: 159), I shall base the following discussion (almost) exclusively on material from the latter category.

Leading articles

Let us now turn to the canonical type of comment text, the leading article. Most of us would probably agree that a prototypical leading article is a text that describes a non-trivial problem, typically political, suggests one or more possible solutions, and weighs their relative merits in the light of their possible consequences. The prototypical leading article, in other words, will be persuasive or expository, depending on whether it contains direct appeals to adopt the solution propagated. However, before proceding to explore such macro-structural and macro-generic features of leading articles, I wish to open this section with some observations on their characteristics at a lower level, namely at illocution level.

Illocution types

As mentioned above, the central task of the news report is to convey information on events that happened in the (recent) past. This kind of information is typically conveyed in the illocution type which Searle (1976) called *representative*, that is, a speech act which commits the speaker to the truth of the proposition expressed. This is the type we find in such newspaper sentences as the following

(1) Assets belonging to five Jewish victims of Nazi Germany have been found in an initial search of Swiss bank vaults. (DTG, 13.11.96: 14)

(2) The Archbishop of Canterbury issued a pre-election warning to clergy yesterday of the dangers of claiming special authority for their political and economic opinions. (T, 15.11.96: 8)

These sentences exhibit at least the following three characteristics of representatives: in the first place they are declarative sentences, second, the verbal tenses employed refer to some time in the past, and third, the claims they make are *factual*, i.e. they are claims concerning events and states of affairs about whose truth there can be objective knowledge. But, as Atelsek (1981) points out, there are also claims about which there can never be objective knowledge, and which therefore do not really fit into any of the categories of speech act provided for in Searle's (1976) classification. The crucial point is that it does not really make sense to talk about the degree to which someone is committed to the truth of a proposition if, strictly speaking, the question of truth is irrelevant.[2] Atelsek, who refers to such claims as *normative*, distinguishes two subtypes: evaluations ("The food was delicious") and proposals ("You ought to learn to cook"). Normative claims are not at all uncommon in news reports, but when they occur, they are normally attributed to a named source:

(3) W N Deshmuki [...] said he believed safety standards in India were the worst in the world. (DTG, 13.11.96: 1)

(4) The Church should not confuse theological and moral pronouncements with politics, Dr Carey said. (T, 15.11.96: 8)

Evaluations and proposals, however, are not the only claims about which there cannot be certainty. In addition, at least the following exist: predictions, causal explanations, and interpretations. Like evaluations and proposals, these illocutions all rest on some sort of human assessment rather than on empirical evidence, and in what follows I shall refer to all five of them collectively as *Assessives*. When they occur in news articles, all these illocution types are typically attributed to a source. In leading articles, on the other hand, they abound and, crucially, without being attributed to sources other than the writer of the text. On the contrary, the source of the assessment is often expressly claimed to be the writer, typically in the shape of the "editorial we", a phenomenon I shall refer to as *self-attribution*. I will substantiate these claims with illustrations taken from the leading articles of the *Daily Telegraph* and the *Guardian* from 19 May 1993 (cf. appendix 1-2).

Evaluations. Unlike representatives, evaluations are about whether things, real or imagined, are good or bad, desirable or undesirable:

(5)   But what matters most, in our view, is how Euro-enthusiasts respond to yesterday's vote. (DTL:12-13)

Note the self-attributive "editorial we".

Proposals. These utterances go one step further than evaluations in that they explicitly point out what needs to be done if the present state of affairs is not satisfactory:

(6)   A healthy Western Europe is important and greater unity is still worth striving for. (G:16-17)

They are arguably a special case of the speech act category "Directive".

Predictions. Predictions can be real, hypothetical or even counterfactual, i.e. based on conditions known to be false:

(7)   This is the wider lens through which the EC will be viewed in the year 2000, whether or not the fine print of Maastricht has been observed. (G:20-22)

(8)   [Euro-enthusiasts] might declare that the Maastricht Treaty is now home and dry [...]. (DTL:13-14))

(9)   [...] a second Danish No would have kiboshed the treaty, the British bill and probably John Major's premiership too. (G:26-27)

Note that although a prediction may well in due course turn out to have been true, it is still, at the time it is made, ultimately based on human assessment.

Causal Explanations. In the physical world we can observe that one event precedes another, but whether or not that precedence relation is also a causal relation is, in the last resort, a question of assessment. When it comes to questions of human actions, assessments of cause and effect are even more precarious, and in many cases, such as the following, causal explanations may well be felt to be verging on innuendo by those whose actions are explained:

(10) A pounds 53 million tax rebate, dangled as the reward for a Yes, also played a part [in making the Danes vote Yes]. (DTL, 19.05.93)[3]

Interpretations. To interpret a phenomenon is to ascribe meaning to it, typically by stating that the "real" meaning of an observable phenomenon is perhaps not the most readily inferrable one:

(11) To declare that [British Eurosceptics] are out of line with the rest of Europe [...] is to misread the signals. (DTL:24-26)

A quick count of the illocutions in the two leaders will substantiate the claim that assessives make up a very important illocution type in leaders: In the *Guardian*, 27 out of 35 illocutions are assessive, and in the *Daily Telegraph* the corresponding figures are 21 out of 27.

To say that the illocutions typical of leaders are illocutions about which there can in principle never be absolute certainty is not to say that the claims propounded in leaders are totally unfounded. Leaders, like all other kinds of discourse about phenomena about which there cannot be certainty, resort to reasoning or argumentation, which is often expressly marked by conjunctions such as *because* in the following:

(12) Last night's Danish Yes is better than a No - if only because the EC cannot afford another self-absorbed year of treading water. (G)

The argumentative structures of leaders are beyond the scope of this paper, however (for further discussion, see Vestergaard 1989, 1995, 1996).

Macro-generic features of leaders

I said above that the leader page of an upmarket newspaper is the place where the paper and its readers talk to each other. That observation must now be supplemented with the further observation that many political leaders in upmarket newspapers will have a dual readership consisting of both lay readers and politicians. In considerations about the communicative purpose - and hence the genre - of a leading article, this dual readership must be kept in mind, for there may well be cases where the communicative purpose of a given text will differ between the two readership groups. The DTL leader from 19 May, 1993 (appendix 1) is a case in point.

If we read the DTL text as addressed to lay readers, this is a persuasive text in almost canonical form: It contains the four "moves" Problem, Solution, Argumentation and Appeal in that order; there is identity between moves and textual segments (although the actual typographical paragraphing is not a very reliable clue to the structure of the text);[4] and finally, there are no cases of embedding or recursion to complicate the pure persuasive structure. Section I (ll. 1-19) describes the *Problem* created by the Danish *Yes* vote to the Maastricht treaty: overenthusiastic pro-European politicians might take it as a go-ahead signal; section II (ll. 20-27) now proposes a *Solution:* we should show the Eurosceptic minorities consideration; section III (ll. 27-37) contains the *Arguments* for the solution: a) the Eurosceptic MPs represent a substantial proportion of the people, and b) a failure to consider them might end in rampant nationalism; finally section IV (ll. 38-44) contains an *Appeal* of the scare-tactics variety: "If we press ahead with integration, we risk creating a nationalist backlash like the National Front in France." The text can thus function both as a warning to Euro-enthusiasts and a comfort to Euro-sceptics.

It is also possible to see the text as addressed to the pro-European majority of Conservative MPs or possibly even Cabinet members, however. Considering that the central difference between persuasive and hortatory texts is that whereas the former tries to boost a point of view, the latter attempts to instigate action, the text must now be understood as hortatory rather than persuasive, in so far as only the politicians in question had the power to act (*in casu*, to refrain from acting) as recommended in the text.

The *Guardian's* leader from the same day on the same topic (appendix 2) is quite different. In the first place there is no question of a dual readership consisting of lay readers and politicians respectively, as the *Guardian*, presumably, did not carry much weight among the higher echelons of the Conservative party, which was then in government; second, there is no evidence for seeing the text as addressed to a readership divided in their attitudes to the problem under discussion: the newspaper is unambiguously pro-European and assumes its readers to share that attitude. The leader is thus an expository text, but quite remarkably, almost the entire text (ll. 1-42) is taken up by the *Problem* move: (a) in spite of the Danish Yes, the goals of Maastricht are less likely to be attained than they were two years ago, and (b) although Britain is sure to follow, British politicians have made themselves the goal of popular contempt. Then follows a brief *Solution* move in which the text proposes a reaction to the problem: there is no reason to be overjoyed; and an equally brief *supporting Argumentation:* for this is not the confident step forward that we could have wished.

The dual readership that we observed above in our analysis of the DTL leader can be played upon deliberately as in the *Sun's* "Andy freebie" (appendix 3), whose closing line is "You owe us three grand, Andy." If the addressee of the text is taken to be Prince Andrew himself, this is a hortatory text, although its *Command* move ("pay us back") is not stated explicitly, and would have to be inferred.[5] In actual fact, of course, the addressee is not the Prince but the paper's lay readers, and in the final analysis the text must be regarded as expository (sic), although it only contains one of the moves characterizing that genre: description of problem.

In addition to expository and persuasive texts, texts that are directly and genuinely hortatory also occur as leaders. In the *News of the World's* "Let's never forget" (Appendix 4) we find an injunction to commemorate the armistice day of the First World War. This brief text manages to comprise all four moves of a hortatory text (cf. above), although not in strict canonical order. First the text presents the *Problem*: the slaughter of World War I ended 78 years ago (ll. 1-5); then the text tries to establish *Authority* by referring to the Queen's participation in the commemoration (ll. 6-9); the next two lines (10-11) try to create *Motivation* by referring to feelings of sorrow and gratitude; then follows a paragraph whose purpose must be understood as another attempt to establish *Authority*: many thousands will commemorate; and finally we get a direct *Command*: "join them."

Above, p. 97, I mentioned how advertisements will be interpreted as advertisements even in cases where they borrow the outward appearances of some other genre (most frequently strip cartoons or editorial articles). This phenomenon, which I propose to refer to as "genre borrowing", also occurs in leaders. Consider appendix 5, "An agonised aunt", the *Daily Telegraph's* third leader on 13 November, 1996. This text was published on the occasion of the death of Marje Proops, the editor of the *Daily Mirror's* agony column (a so-called agony aunt). It is couched in the question - answer format of the agony columns of the popular press, the question ostensibly being written by a certain *M.*, who regrets having given many people too liberal advice and cared too little about traditional morality. But since the text is printed where it is, it will of course get the interpretation of a leader, i.e. it will be read as presenting a Problem (today's libertarian attitudes in matters of sex and marriage) and proposing a Solution (pay more heed to traditional values). From a theoretical point of view, what is interesting about genre borrowing is that it provides compelling arguments for regarding genre as a purely social phenomenon, i.e. in relation to its function in the discourse community. For in the case we have just been considering, the text became assigned to a certain genre solely on functional grounds (here on the basis of the context in which it appeared) and without the least bit of intra-textual evidence. We shall explore this topic further in the next section.

Other non-news genres

Not every text printed in the news pages of an upmarket newspaper is news, and not all comments and analyses are restricted to the "comments" pages. A modern (upmarket) newspaper consists of a number of special sections (finance, arts, sports, etc.) in addition to the "hard news" section, the section in which we find reports on "accidents, conflicts, crimes [...] which have come to light since the previous issue of the paper or programme." (Bell 1991: 14), and in fact, it is only in these pages that papers even purport to observe the distinction between reporting and commenting. For as a quick examination will show, the distinction is blurred or simply irrelevant in all the various special sections of a modern newspaper: in the finance and business sections reporting and analysing merge imperceptibly, in the arts pages reporting and evaluating (reviewing) are inseparable, and in the sports pages much more evaluation is allowed to creep into a report from e.g. a football match than would have been tolerated in a piece of politcal news reporting. But even in these sections, the report - comment distinction is maintained, at least symbolically, in that they will carry comments written by special columnists and bearing their names and photos much as in the hard news sections. A final genre in which, for obvious reasons, factual reporting and evaluation merge is the obituary, which the upmarket papers, characteristically, print on the leader pages or the immediately preceding or following pages.

However, texts that are explicitly non-news also occur in the hard-news sections of serious newspapers, consider for example appendix 6.a.-b. (the *Times,* 15.11.96:3). The text in appendix. 6.a. has many of the features of the typical hard-news story: it relates a sequence of events having taken place "since the previous issue of the paper" (namely "yesterday"), and in the standard manner of newspaper reporting, the temporal sequence of the narrative is cyclical rather than linear (cf. van Dijk 1988, Bell 1991). Over half of its finite verbs are in the simple past, and although it contains more unattributed evaluation than we would expect in a serious news story (e.g. "the dubious practice of seeking to shrive herself on the shimmering small screen..." [ll. 43-47]), most of its assessive claims are attributed to sources other than the writers of the news story (e.g. "America, it's time to dump royal pain Fergie" [ll. 28-29], and "[...] bookshops expect that to change [...]" [ll. 144-145]).

The text in appendix 6.b. on the other hand, is clearly marked as being non-news: we are not only given its author's name, but also his title (Dr) and his photo, and although the article occurs on a "home news" page, we are told in so many words that this text belongs in a different genre, as it is a "medical briefing". On a more general level we would say that this is an example of the *"background"* genre, which does not contain news but provides knowledge of various kinds enabling the reader to assess and understand the news better. Note that two thirds of its finite verbs (21 out of 32) are in the simple present tense, and that evaluations and predictions abound and are understood to be the writer's own (e.g. "The duchess of York is unfortunate..." [ll. 1-2], "But no blinding flash...will ever turn her into an astute banker or nun." [ll. 63-67]). There is no temporal succession, so we would be inclined to regard it as an expository text: it certainly describes a problem, but there is no proposed solution nor any evaluation of the solution. Instead it contains a diagnosis and most of the article can be regarded as argumentation for the correctness of the diagnosis.

The two texts in appendix 7.a.-b. again exemplify a news story proper accompanied by a background article (this time from the *Guardian*, 15.11.96:1). Note that, as in the previous case, the background article is marked typographically as different from the rest of the page. Linguistically, what sets it off from real news is not that it contains evaluations, predictions, interpretations, etc., for it does not. What does characterize it as non-news is that it is not narrative. Roughly half of its finite verb phrases are in the simple past (including past passives) to be sure, viz. 11 out of 24, but it simply lists a series of facts, past or present, which are capable of throwing light on the miserliness exhibited by the behaviour of the main character of the events narrated in the main news story. These facts, however, are not presented as being related sequentially.

We are here faced with two texts which are clearly non-news in Bell's (1991) sense of the term (cf. above) in that they are not reports of events that have happened since the publication of the last issue of the paper, and which we would intuitively like to group in a common genre, "background", but which purely intra-textual considerations would force us to regard as examples of two different genres, Expository and Descriptive: they differ from each other in several respects, ranging from micro-level features such as verb tenses (present vs. present and past), over their use of illocution types (predictive and evaluative vs. factual) to the macro-level, where one argues a case, and the other lists a series of facts. It is only when we consider the two texts in their textual and social contexts that we are able to say that they are of "the same kind": both are placed in the immediate vicinity of another text, both deal with the same subject matter as that other text, and both put readers in a better position to appreciate and assess the information given in the main text.

Conclusion

The leading article, as we have seen, is centrally realized by Hortatory, Persuasive or Expository texts. However, as defined in this paper they all have so much in common that it is not difficult to see them as forming one macro text type: They all deal with problem situations, and they all present solutions to the problems dealt with. The main difference between them is the degree of urgency with which they call upon addressees to accept, and possibly even act on, the solutions proposed, and in fact it is possible to see them not as discrete types, but as points on a continuum. Here Expository texts, which restrict themselves to proposing solutions and arguing for them but refrain from urging the addressee to accept them, clearly belong at the least urgent end, and Hortatory texts, which not only, more or less openly, command the reader to follow their instructions but on top of that may even claim authority to do so, belong at the most urgent end, with Persuasive texts falling between the two. At the micro-level, too, they have much in common: all of them provide evaluations, give interpretations, ascribe causes, make predictions, recommend courses of action, i.e. all of them are assessive. If we disregard the relatively rare cases of genre borrowing such as "An agonised aunt", discussed above, the typical situation is for there to be a quite straightforward relation between the leader genre and the text types realizing it.

It is when we come across examples such as the two background texts in appendix (6.b.) and (7.b.), that the idea of a simple one-to-one relation between genre and text type is really challenged. If we were to rely on purely text internal evidence, there would be little reason why anyone would wish to regard them as members of the same class, for they differ from each other at every level of analysis. At the syntactic micro-level in their employment of verb tenses; at illocution level in their use of predictive and evaluative judgments in the one case and factual statements in the other; and at discourse level in that one argues a point of view, whereas the other lists a series of facts. Yet, if we consider the extratextual evidence, both texts have much in common. Both are placed immediately adjoining another (main) text, both deal with the same subject matter as that other text, and both provide readers with information that will enable them to arrive at a better understanding of the main text. It is for this reason that we might well wish to classify them under a common generic label, "background". Understood in this way, "genre" would be a concept that is based not just on text internal criteria, but also, and crucially, involves functional considerations. I am not in any way suggesting that I am at odds with commonly accepted genre theory in making these claims, since the pivotal concepts of "communicative purpose" (Swales, 1990; Bhatia 1993) or achieving "different culturally established tasks" (Eggins and Martin, 1997) themselves are text-external phenomena. What I am saying is that there are cases where purely text-internal considerations of the text itself do not offer us sufficient evidence for deciding upon the communicative purpose, and hence the genre, of the text in hand. If this is accepted, and if it is accepted that there may be reason to regard texts as diverse as the two "background" articles above as representatives of the same genre, there will no doubt be a need for a distinction between classifications based on text external and text internal criteria (cf. also Pilegaard and Frandsen 1996, Trosborg 1997). We could then reserve the term "text type" for text internal categorizations, and it would be possible for two texts to represent the same genres but different text types.

Abbreviations

DTL   the *Daily Telegraph*
G   the *Guardian*
T   the *Times*

Notes

[1] Although there are evident similarities between Longacre's four-move structures for e.g. persuasive and expository texts and those proposed by Bhatia (1993: 165-168), on the whole, "move", in Bhatia's conception of the term, seems to be a lot closer to an element in actual textual sequence. It should be noted, however, that Bhatia does point out that moves are not identical with paragraphs (1993: 56), and that one move may be "syntactically embedded" in another (op.cit.: 89).

[2] Note that I am here talking only of *claims* about whose truth there can never be certainty. There are of course several other types of illocutionary act in which the question of truth is irrelevant, viz. Searle's Commissives (e.g. promising), Directives (e.g. requesting), Expressives (e.g. apologizing) and Declarations (e.g. authorizing).

[3] This example is not from a leader but from a feature article written by a named columnist. Causal ascriptions do not occur in the two leaders examined.

[4] For increased readability, newspapers tend to use indentation far more frequently than warranted by the content structure of their texts.

[5] This observation contradicts Longacre (1992) and Eggins and Martin (1997), who agree that the Command move or stage is obligatory in hortatory texts.

References

Atelsek, J. 1981. "An anatomy of opinions". *Language in Society* 10: 217-225.
Bell, A. 1991. *The Language of News Media*. Oxford: Blackwell.
Bennett, T. 1982. "Media, 'reality', signification". In M. Gurevitch et al. (eds), *Culture, Society and the Media*. London and New York: Methuen, 287-308.
Bex, T. 1996. *Variety in Written English*. London: Routledge.
Bhatia, V.K. 1993. *Analysing Genre: Language Issues in Professional Settings*. London: Longman.
Coulthard, M. 1994. "On analysing and evaluating written text". In M. Coulthard (ed)

1994, 1-11.
Coulthard, M. (ed) 1994. *Advances in Written Text Analysis*. London and New York: Routledge.
van Dijk, T.A. 1988. *News as Discourse*. Hillsdale, N.J.: Lawrence Earlbaum Associates.
Eggins, Suzanne and Martin, J.R. 1997. "Genres and registers of discourse." In T.A. van Dijk (ed), *Discourse as Structure and Process*. London: Sage, 230-255.
Fowler, R. 1991. *Language in the News*. London: Routledge.
Fowler, R. et al. 1979. *Language and Control*. London: Routledge & Kegan Paul.
Halliday, M.A.K. and Hasan, R. 1976. *Cohesion in English*. London: Longman.
Halliday, M.A.K. and Hasan, R. 1985/1989. *Language, Context and Text: Aspects of Language in a Social-Semiotic Perspective*. Oxford: Oxford University Press.
Hartley, J. 1982. *Understanding News*. London and New York: Routledge.
Hodge, B. 1979. "Newspapers and communities". In R. Fowler et al., 1979, 157-174.
Hoey, M. 1994. "Signalling in discourse: a functional analysis of a common discourse pattern in written and spoken English". In M. Coulthard (ed) 1994, 26-45.
Katz, J. 1987. "What makes crime 'news'?" *Media, Culture and Society* 9: 47-75.
Longacre, R.E. 1972. *An Anatomy of Speech Notions*. Lisse: Peter de Ridder.
Longacre, R.E. 1992. "The discourse strategy of an appeals letter". In W.C. Mann and S. A. Thompson (eds), 109-130.
Mann, W.C., Matthiesen, C.M.I.M. and Thompson, S.A. 1992. "Rhetorical Structure Theory and text analysis". In W.C. Mann and Thompson (eds), 39-78.
Mann, W.C. and Thompson, S.A. (eds) 1992. *Discourse Description. Diverse Linguistic Analyses of a Fund-Raising Text*. Amsterdam/Philadelphia: John Benjamins.
Martin, J. 1992. *English Text*. Amsterdam/Philadelphia: John Benjamins.
Pilegaard, M. and F. Frandsen. 1996. "Text type". In J. Verschueren et al. (eds), *Handbook of Pragmatics 1996*.
Searle, J.R. 1976. "The classification of illocutionary acts". *Language in Society* 5: 1-24.
Stubbs, M. 1996. *Text and Corpus Linguistics*. Oxford: Blackwell.
Swales, J.M. 1990. *Genre Analysis: English in Academic and Research Settings*. Cambridge: Cambridge University Press.
Tannen, D. 1979. "What's in a frame? Surface evidence for underlying expectations". In R.O. Freedle (ed), *New Directions in Discourse Processing*. Norwood, N.J.: Ablex, 137-181.
Trosborg, A. 1997. "Register, genre and text type". In A. Trosborg (ed), *Text Typology and Translation*. Amsterdam/Philadelphia: John Benjamins Publishing Company.
Vestergaard, T. 1989. "Assertions and assumptions in argument". *English Language Research Journal* 3: 43-57 (English Language Research. University of Birmingham).
Vestergaard, T. 1995. "On the open-endedness of argument". In S. Millar and J. Mey (eds), *Form and Function in Language*. Odense: Odense University Press, 137-150.

Vestergaard, T. 1996. "Argumentationsmønstre i britiske avisledere". In F. Frandsen (ed), *Medierne og Sproget.* Aalborg: Aalborg Universitetsforlag, 33-50.
Vestergaard, T. and Schrøder, K. 1985. *The Language of Advertising.* Oxford: Blackwell.
White, P. 1997. "Death, disruption and the moral order: the narrative impulse in mass-media 'hard news' reporting". In F. Christie and J.R. Martin (eds), *Genre and Institutions. Social Processes in the Workplace and School.* London: Cassel, 101-133.
Wilsher, P., Macintyre, D. and Jones, M. 1985. *Strike: Thatcher, Scargill and the Miners.* London: Hodder and Stoughton.

Appendix 1

DTL 19 MAY 93 / Leading Article: Yes, but

ONE IS enough, Churchill once said of majorities, and the Danes on second thoughts have done better than that. Given the choice of restoring much needed confidence in the Maastricht Treaty or virtually wrecking it, they have stepped back from the brink.

5 Among ministers in London there will be profound, if cautious, relief, reflected by the Prime Minister's speech to the Confederation of British Industry last night. Brussels will sound a note of subdued rejoicing. And the minority of Tory Euro-sceptics will declare their intention of continuing the fight by one means or another more intensely than ever. Those
10 campaigning for a referendum in this country will regard the Danish result as strengthening their hand.

But what matters most, in our view, is how Euro-enthusiasts respond to yesterday's vote. They might declare that the Maastricht Treaty is now home and dry, and interpret this second Danish vote as a signal to go full steam
15 ahead towards ever closer European union. They might regard any more foot-dragging in the British Parliament as wholly irrelevant. Some may even take their cue from Sir Edward Heath and dismiss the foot-draggers as virtually enemies of their country - an extraordinary view of parliamentary democracy. But such a response, in our view, would do great harm.

20 What needs to be heeded, in the wider interests of Europe, is the existence in this country and the other 11 members of the Community of substantial minorities of citizens who simply do not want to live in a European federation founded on extremely flimsy ideas of democratic accountability. A high proportion of this Government's supporters are of that mind. To declare
25 that they are out of line with the rest of Europe, as do people like Martin Bangemann, German vice-president of the EC's executive commission, is to misread the signals. This handful of despised Euro-sceptics in the House of Commons, these non-citizens, represent a much higher proportion of Conservative Euro-sceptics outside Parliament.

30 If the substantial minorities in Denmark, Germany and France, as well as Britain, are now brushed aside as if their views do not matter, there will slowly gather a storm in Europe. Churchill's aphorism about one vote is not wholly applicable here.

The debate is about fundamental and permanent constitutional changes, for
35 which some nations require a two-thirds majority. One danger is that such a storm would carry Europe not towards closer co-operation but in the opposite direction: towards rampant nationalism.

We should take note of the fact that the French National Front won an eighth of the vote in the recent general election. Nationalism is a latent force.
40 It would feed eagerly on the anxieties of those alarmed by the declared ambitions of people like Martin Bangemann, who seeks 'an integrated political system' for Europe based on a federal structure. If Europe is to move forward peacefully, it must pay more attention to the discontent. This Danish vote is not a call to go faster, but to be more heedful.

The Daily Telegraph

Appendix 2

SOURCE: The Guardian DATE: 19 May 1993 FEA PAGE: 19
Leading article: Yes, but: and the real road forward

LAST night's Danish Yes is better than a No - if only because the EC cannot afford another self-absorbed year of treading water. But there is little point in pretending that life has not changed in as many ways as the map since European union was first devised. You only have to look at the future laid down in the Maastricht Treaty to see that it no longer speaks to the realities of the continent. A single currency by the end of the century and economic policy largely in the hands of a European Central Bank? With two opt-outs already registered, and a Germany in ever-deepening recession, that is an obvious area for slippage. A common foreign policy to ensure that EC members "speak with one voice"? As Yugoslavia has already shown, this is a massive cart to put before the pantomime horse of European disunity. A common defence policy? The same problem with nuclear additions. Cooperation on justice and home affairs is more promising (though Bosnia casts a shadow on the right of political asylum). But the setting up of Europol and an EC-wide system for exchanging information is not exactly at the heart of Maastricht. How it meshes with an enlarged community is much less certain than it was two years ago.
A healthy Western Europe is important and greater unity is still absolutely worth striving for. But the argument is already less about making new gains, more about how to defend this core against rising unemployment, falling investment, and a loss of social identity.
The EC can claim no hegemonic role towards the larger Europe which concerns us most of all. It is part of a common effort now centred on the UN. This is the wider lens through which the EC will be viewed in the year 2000, whether or not the fine print of Maastricht has been observed.
Meanwhile, the Danes have saved Britain's bacon. No other nation in Europe had as much at stake in the Danish vote. Denmark and Britain, respectively the rejectionist and the reluctant, have agonised too long. As Bill Cash and his chums well knew, a second Danish No would have kiboshed the treaty, the British bill and probably John Major's premiership too. Now that the rebels' great hope is undone, all three live to fight another chastened day. The sting has been drawn from Thursday's third reading debate in the Commons and from the remaining stages of the British ratification process. Though a series of noisy debates in the Lords and the subsequent Commons vote on the social contract still loom, they loom smaller. Legal hurdles likewise remain, though with less menace. The long night is over.
As in other countries of the twelve, the British political class's absorption in Maastricht has cost it dear. While MPs have been engrossed in an epic and, to them, wonderful parliamentary battle, their reputation in the world beyond Westminster has gone on the slide. Politicians have rarely been so despised, a constant theme in Newbury and the shires. The problem is not that Maastricht was or is unimportant. On the contrary. Yet parliament and its procedures seemed always to blur the big issues at the heart of the debate, not to clarify them. The sense of a nation taking an historic decision was constantly mocked by the obscurity of a Westminster process which has been unable to rise to the occasion. The Danish decision has been taken and Britain will now come ruefully into line. But this is no time for grand rhetoric. This is not so much a confident stride into the future as a hesitant shuffle along a wavering line into much remaining uncertainty.

(c) 1993. All data on the CD-ROM is in copyright.
Contact Guardian Newspapers Ltd for permission to reproduce

Appendix 3

Andy freebie

PRINCE Andrew has got a cheek.

There was no need for him to waste £3,000 of taxpayers' money on a helicopter jaunt to his daughters' school. He could easily have driven the 20 miles — in his own car, using his own petrol.

We thought the royal freebies had ended with Fergie.

You owe us three grand, Andy.

Appendix 4

Let's never forget

IT is 78 years since the Armistice which, at 11 o'clock on November 11, 1918, ended the terrible slaughter that would become known as the First World War.

Today, led by Her Majesty, the nation pays its respects to all those who have died in the conflicts, then and since.

Despite the passing years, the feeling of sorrow and gratitude is still strong.

And, as if to emphasise that this is no routine lip service to remembrance, many thousands will tomorrow observe another two minutes' silence—at the 11th hour of the 11th day of the 11th month.

Show you care—and join them.

Appendix 5

An agonised aunt

Q. CAN you help me? I am an elderly widow who recently passed away. Over the years I gave advice to millions of people on how to live their lives. I know many of them appreciated it, and I'm proud of that. But now that I've got more time to think, I wonder if I always did the right thing. I said they should change the laws to allow homosexuality, abortion and easier divorce. I said people should do what they really wanted, and not worry about morality and religion and all that stuff. I reckoned it was better to fulfil yourself, even if it sometimes meant other people getting hurt. Now that there's so much family breakdown and violence and child abuse, I'm not so sure. From where I sit today, some of those old teachings don't look so stupid. So please can someone tell me, where did I go wrong?
M. (name and address withheld)

A. Don't worry, love. Most of us made the same mistakes. When you were young, the rules seemed too harsh. You wanted to break out, and who can blame you? You thought that, if only people were more honest, we could all build a better world and love one another more. You saw that women got kicked around, and you didn't like it, quite rightly. A lot of what you said needed saying. It turned out to be a bit of a case of the baby being thrown out with the bathwater, that's all. Those of us who are left down here will just have to try to sort it all out somehow. But if you're where I think you are now, your Dad will know that you did what you thought was right, and that'll be good enough for Him. And if you're where I hope you're not, no aunt can help your agony.

Why not send us some more messages from the other side, Marje? After all, you must be able to get the real truth now. On second thoughts, don't. You'd put all of us out of a job.

Appendix 6a+6b

6a

America sticks the knife into the Duchess of Talk

BY QUENTIN LETTS IN NEW YORK AND EMMA WILKINS

THE Duchess of York hit a cold Manhattan yesterday on the latest leg of a self-publicity tour, only to be told that she was "Britain's most unwelcome export since mad cow disease".

It was not only winter's first snow flurries that chilled the air. The local media suddenly wearied of her, failing to show any gratitude for the duchess's statement this week that she loves America and wants to live there.

The duchess, who is touring the United States to promote a book and rebuild her reputation, wore a skirt slit to the thigh as she swept into a Fifth Avenue bookshop to sign copies of her children's story. At the same time, a television network was debating "Is Fergie a royal pain?" and the New York Post was carrying a vitriolic attack headlined "America, it's time to dump royal pain Fergie".

"Someone ought to tell Fergie that America no longer needs foreign royalty," it thundered. "Change the Channel. Don't buy her tell-all book."

The Duke of York's former wife crossed the Atlantic apparently believing that it would do her good to adopt the American habits of public self-analysis and soul-baring. She has appeared before the grand queens of broadcasting, Oprah Winfrey and Diane Sawyer and engaged in the dubious practice of seeking to shrive herself on the shimmering small screen to tens of millions of Americans. Yesterday's barrage suggested that the scheme was a failure.

Her friend-turned-enemy, Allan Starkie, was in New York at the same time, hawking his sordid version of the Fergie story. Dr Starkie, from Long Island, offered poisonous indiscretions that threatened to tarnish further the duchess's name.

None of it boded well for the final negotiations she is conducting with Weight Watchers, the dieting organisation, for which she hopes to become a front woman in exchange for $1 million.

The duchess's desire to present herself as an ordinary Joan with human failings have made it an embarrassment to be a Briton abroad. Interrogated by Diane Sawyer on the nationwide ABC network, the duchess threw her face into contortions of suffering as she described her life at Buckingham Palace.

Throughout her meanderings she has praised the conduct of the Duke of York. For all the good intentions, however, the result has still been negative. Reporters concentrated merely on her denial that the Duke was homosexual, and portrayed the Queen's second son as little more than a dullard who watches too much television.

Amid the ghastly circus, however, one possible boon presented itself yesterday The New York Post polemicist Andrea Peyser, made such a robust attack on "former royals who would sell out Buckingham Palace for a buck" that it is possible the monarchy may yet benefit, simply by being the betrayed institution.

In Britain, in an interview on Radio 4's Today programme yesterday, the duchess pointedly avoided a question about her fidelity to her husband during their marriage. She said that the issue was "not relevant" to the interview.

She did choose, however, to declare her intention to repay debts estimated at £4 million to Coutts, her bankers. Comparing her addiction to spending money with her sister-in-law's bulimia, the duchess said she had now conquered her desire to lead a lavish lifestyle.

Motherhood was probably her only success, she said. "The only thing I can probably say I am good at is being a mother," she told Sue McGregor, her interviewer.

The duchess said that after the embarrassing publication of photographs of her toes being sucked by John Bryan, it was only her religious faith that sustained her.

Brushing aside suggestions that she had received "dressing-downs" from the Queen, the duchess said: "I would like to think that Her Majesty is grandmother to my two, to our two children, and therefore we carry on that relationship."

Her book, meanwhile, was being outsold by Dr Starkie's account even in Britain. But bookshops expect that to change as the publicity blitz takes effect.

My 'Story: the 'tell-all book by the Duchess

6b

The temptations she cannot resist

THE Duchess of York is unfortunate in that her personality is likely to lead her all her life into social, financial and sexual scrapes of varying magnitude. They probably stem from a condition which is labelled "failure of impulse control".

During her interviews yesterday the duchess expressed regret for the difficulties she has caused others, but classically seemed to be emotionally detached. The magnitude of the disasters she had occasioned did not seem to be causing severe anxiety.

Her present preoccupations not unnaturally centre on her debts, which she says have slipped all too easily

Dr Thomas Stuttaford

from six to seven figures. The size of the debt can be attributed to her buying and spending sprees; they are not so much a matter of immaturity, as some people display that trait all their lives.

In other people, the same personality defect of failure of impulse control leads to pathological gambling, constant travelling, or even pyromania. Whatever the desire, the feature of the disorder is that patients cannot resist an impulse. Naturally, the consequences of their act, whether its a debt at Coutts or a gambling loss in the casino, causes regret and remorse, but sooner or later, loneliness, depression or unhappiness will trigger another bout of buying or spending, and the cycle is repeated.

Disorders of impulse control are often very much part of a wider picture and of difficulties with personal relationships. As with any personality disorder, treatment is difficult. As the duchess herself says, her troubles may well have been laid down in childhood when her home life was fractured.

The duchess's principal concern now is for her children, and there is every reason to suppose that she will be a devoted mother. But no blinding flash on the road to Damascus will ever turn her into an astute banker or nun.

Appendix 7a+7b

7a

£1,200 payout too little for me, says duke

Stuart Millar

HE is worth £5 million, owns 20,000 acres and can boast that his second home is regarded as the most romantic house in England. But His Grace the Duke of Rutland, Britain's 54th wealthiest man, is locked in battle with a building society over the regal sum of £1,200.

The normally reticent 77-year-old Duke, whose art treasures alone are valued at £34 million, has provoked fury from small investors in the Alliance & Leicester with demands that he should receive a greater share of the £50 million the society is offering to members when it becomes a bank next year.

He and 1,000 other wealthy members — with investments totalling £50 million — dislike the A&L's across-the-board policy of a vote and bonus share offer that will give all 2.4 million members a windfall of about £1,200 each

as "unfair and unjust". They claim the offer should be graded.

"They are threatening to close their accounts and vote against the plan when it is put to investors next month.

"If investors with a few thousand pounds were to take their money out, and put it somewhere else, it would cause embarrassment to the society," the duke said from Belvoir Castle, his magnificent home in Leicestershire. He said he and 1,000 other young savers of the society's reputation and "out of local loyalty".

He insisted he spoke for both small and large investors, and was simply against "carpetbaggers" making a profit.

"They have the right to do what they want but it is inequitable. If you have been an investor and put quite a bit of money into a building society and then you see someone who put their money in a couple of years ago get the same bonus, it is unfair and unjust."

The bonus offers were attacked last week by relatives of severely disabled people who will not receive any shares because they rely on other people to administer their accounts.

The A&L yesterday refused to bow to the duke's demands. A spokesman said: "It might not seem a lot to somebody who is worth this much, as he is, but to the vast majority of our members, who have less than £1,200 in their accounts, it means a lot."

He said some investors had written expressing their anger at the duke's intervention. "Perhaps he would like to explain to these customers, such as the Oxfam pensioners, who exists some of their money."

7b

Multi-millionaire with eye for a bargain, Alliance and Leicester building society's across-the-board share offer "unfair and unjust" to wealthy members

The Duke and Duchess of Rutland: Alliance and Leicester building society's across-the-board share offer "unfair and unjust" to wealthy members

BORN Charles John Robert Manners in 1919, the 10th Duke of Rutland inherited his title and his fortune at the age of 20 when his father died, writes *Stuart Millar*.

Last year, he was valued at £78 million — £1 million poorer than pop singer Paul McCartney.

Belvoir Castle, the family seat, boasts 356 rooms and sits in 18,000 acres of land. The estate is also home to the Duke's foxhounds, the Belvoir Hunt, where Prince Charles has expressed her distaste for field sports.

But it is the Manners' second home that attracts most admiration. Haddon Hall, in Derbyshire, which dates from the 13th century and came into the family through marriage in the 1500s, has been described as the most romantic house in England. Restored by the duke's father after years of decline, it has acres of terraced gardens and a trout lake.

After a spell as a captain in the Grenadier Guards, the duke consolidated his fortune by exploiting the minerals and as proprietor

of the leisure group Rutland Hotels Ltd.

In 1968, he underlined his business acumen, cornering the market in early Christmas holidays by offering Yuletide breaks at Belvoir over the August bank holiday.

A lifelong Tory, he is as afraid of losing out under Labour as through the Alliance & Leicester. Last year, he presided over a meeting of aristocrats who were briefed by a team of accountants on how to take forward tax preparations against politicians.

The duke is known as much for his looks as his wealth. He was once seen as a suitable escort for the then Princess Elizabeth, and is widely recognised to be the inspiration for many of Barbara Cartland's bodice-ripping heroes. Even now she describes him as the "most handsome man in England".

A keen angler, he is well used to making enemies. Last summer, 8,000 trout in his lake were poisoned with cyanide, and threats have been made against extreme anti-hunting groups.

The Inaugural Address
President Clinton's 1993 address

ANNA TROSBORG
The Aarhus School of Business

This article is concerned with the inaugural address as a genre. It analyses President Clinton's 1993 address in particular and focuses on communicative functions and rhetorical strategies. As a political speech, the address shares many characteristics with political speeches in general, but it also stands out due to the uniqueness of the situation.

A genre analysis typically includes determination of the communicative purpose of the text, the move structure and the rhetorical features employed in constructing the text (see Swales 1990, Bhatia 1993). This analysis is no exception. It starts out, however, analysing communicative functions in order to find out what messages the speaker intends to communicate to his audience, it analyses the rhetorical strategies for conveying these messages and, finally, on the basis of the analysis, it concludes about the communicative purpose and move structure which set off the inaugural address from other political speeches.

Communicative functions

Normally, political speech has the ulterior motive of persuading the audience to vote for the speaker, or the party, or to support a particular policy. An inaugural address is somewhat more subtle; a newly elected president does not need votes right away, so he can tone down that part of his speech.

The topic of President Clinton's inaugural address is the renewal of America and the facts making this renewal difficult. When looking closely at the text of his speech, it is obvious that a variety of communicative functions employing a wide range of speech acts have been applied to formulate his appeal to the American people.

The communicative functions of the speech are analysed within a framework building on Roman Jakobson's theory of language functions (1960), Austin's (1962) and Searle's (1976) classification of speech acts and the later adoptation of these by Tragoutt and Pratt (1980). Central to all of these are the informative/the representative function, the expressive function and the directive function. In addition, Roman Jakobson also established the phatic, the metalinguistic and the poetic function. My framework keeps the phatic function, whereas his metalinguistic function is not central to this analysis. His poetic function is given independent status in the form of rhetorical strategies. It is not treated as one of the communicative functions, as it is seen as stylistic devices rather than as part of communicative action. The classifications presented by Searle, and by Tragoutt and Pratt both comprise representatives, expressives, directives, commissives and declarations; they only differ in so far as Tragoutt and Pratt keep the category verdictives, originally introduced by Austin,[1] as a separate category comprising acts which evaluate and relay judgement (e.g. assessing, estimating, etc.). This category is also employed by the present study.

The phatic function

A necessary condition for communication to take place is to attract attention and ensure that the channel of communication is open. In this case, it is essential for the President to attract the attention of the audience in the first place and to catch people's interest in order to make them listen. Therefore, he starts his address exploiting the phatic communicative function:

> ll. 2-6 "My fellow citizens: To day we celebrate the mystery of American renewal... the vision and courage to reinvent America".

"My fellow citizens" is a direct address to the American population. The use of "fellow" denotes that Clinton wants to create an equal-to-equal social relationship. Participation is complex, as he wants to draw in his audience in a joint effort to reach his goal. This attitude is further underscored by heavy use of the first person plural pronoun *we* (and the corresponding forms *our/us*) to signal that "we all stand together in the fight for renewal". In addition we find a number of inclusive challenges: *Let us do...* is repeated many times to implement his ideas.

The representative function

The representative function is used to present facts about the world: *the direction of fit* is "words to the world", i.e. the sender tries to make his words match the world. Hence representatives are objective statements about the world, which can be verified as true or false. Typical speech acts in this text are reports and informative statements. Clinton offers *reports* on history with use of common historic references which appeal to the American people. Thus, he refers to the declaration of independence:

> ll. 6-9 "When our founders boldly declared America's independence to the new world..."
> l 23 "When George Washington first took oath...".

George Washington was America's first President, and Thomas Jefferson (the third president) wrote the Declaration of Independence in 1776.

Clinton gives *information* about the world of today and of the American people in particular:

> ll. 19-30 "Raised in unrivaled prosperity ... all across the earth". We are told that "Communications and commerce are global, investment is mobile, technology is almost magical; and ambition for a better life is universal" (ll. 27-28).
>
> ll. 33-34 "This new world ...", ll. 48-49 "Americans have ever...", ll. 50-55 "From our Revolution to the Civil War... our people have mustered the determination to construct from these crises the pillars of our history".

Clinton builds up his appeal soliciting common ground. He creates common ground ("pillars of our history") and solidarity by combining past and present. Note that even his representatives are spiced with connotative elements, such as "boldly", "unrivaled prosperity", "magical", "ambition", "mustered", "determination", "pillars of our history", which add an expressive touch to his statements.

The expressive function

In an expressive text the sender's attitude shines through. There is no direction of fit, and expressives are neither true nor false as they are expressions of the sender's psychological state. An expressive text is characterized by being subjective, and most of what Clinton states in his speech is *his* personal opinion. His opinion comes out clearly is his choice of lexis, words which have a connotative meaning (we get associations when hearing/reading them), such as descriptive adjectives and adverbials, loaded verbs and nouns, etc.

Clinton expresses hope that the Americans can change the American society despite the difficulties they confront:

ll.5-6 *A spring reborn* in the *world's oldest democracy*, that *brings forth the vision and courage to reinvent America.*

ll.15-19 [...] a generation of men and women whose *steadfastness* and *sacrifice* triumphed over depression, facism, and Communism. Today, a generation raised in the shadows of the Cold War *assumes new responsibilities in a world warmed by the sunshine of freedom* but threatened still by ancient and new plagues."

ll.30-31 "*Profound and powerful* forces are shaking and remaking our world,..."

ll.47-48 "Though our challenges are fearsome, so are our strengths".

Clinton uses many descriptive adjectives with positive connotations "boldly" (l. 7), "profound" (l. 28), "dramatic" (l. 55), "bold" (l. 61), and nouns like "liberty" (l. 10), "sacrifice" (l. 15), "hatreds (l. 19) and "determination" (l. 52).

He appeals to traditional American Values: the nation, history, hopes for the future, freedom, democracy, renewal, religion, the values of the family, the pride of being an American, etc. (see lexical chains below).

By playing on patriotic feelings and shared values he appeals to the American people and tries to create new visions. Clinton tries to establish himself among the great men (i.e. great presidents) in American history (cf. the many intertextual references: Washington, Jefferson, Kennedy and now Clinton). Kennedy was a young president and so is Clinton. There is hope for "renewal"; through him there is hope for a "new beginning".

The verdictive function

Verdictives, as defined by Austin (1962: 150), are "typified by the giving of a verdict, as the name implies, by a jury, arbitrator, or umpire. But they need not be final; they may be, for example, an estimate, reckoning, or appraisal. It is essentially giving a finding as to something - fact, or value - which is for different reasons hard to be certain about." In a more general sense, verdictive acts evaluate and relay judgement, e.g. assessing, estimating, etc.[2] (cf. Hatim and Mason 1990: 60). The sender passes a judgement on the reciever, a state of affairs, etc. as regards, for example, soundness and unsoundness or fairness and unfairness. Clinton gives his evaluation of the current state of affairs. A striking example is given in ll. 34-47:

"But when most people are working harder for less; when others cannot work at all; when the cost of health care devastates millions and threatens to bankrupt many of our enterprises, great and small; when fear of crime robs law-abiding citizens of their freedom; and when millions of poor children cannot even imagine the lives we are calling them to lead - we have not made change our friend [...] we have not done so. Instead we have drifted, and that drifting has eroded our resource, fractured our economy, and shaken our confidence".

The directive/commissive function

The directive function focuses on the receiver. The sender wants either to get the receiver to do something or to change the receiver's mind. The direction of fit is "world to the words" and the sender wants to commit either the receiver (in directives) or him/herself (in commissives) to future acts. Typical features to look for are

in the case of directives: imperatives, modal verbs (e.g. *may, need, should, ought to, must, have to*), and performative verbs[3] (e.g. *order, request, beg, plead, permit*, etc.; *let us...* encourages a joint effort

in the case of commissives: performative verbs (*promise, offer, pledge, commit, covenant*) and modals expressing commitment (e.g. *will*).

Clinton's expressive language ends in direct appeals to the nation (ll. 61-68). The directive features of his speech are aimed at all Americans, who *must* work together with their President to "renew America". "We must" is repeated six times. The President wants to change the minds of the American people in order to commit them to make a serious contribution to changing the society for the better. He appeals to the people to take responsibility, emphasizing the obligations of the people as well as his own. Later in his talk he puts forward proposals for what to do - *Let us* is repeated a number of times.

Declarations[4]

Declarations require extralinguistic institutions for their performance. It takes a priest to christen a baby or to perform a marriage ceremony in a church, it takes a judge to sentence a defendant, a dignitary to name a ship, etc. The direction of fit is both "words to the world" and "world to the words", as the actual expression of the declaration brings about a change of reality.

Prior to his inaugural address, Clinton has just taken the presidential oath. This oath, which is only referred to in his speech, is part of the inauguration. Taking the oath is part of the ceremy, and by taking the oath the inauguration is performed.

What comes close to a declaration in the unaugural address is Clinton's salutation to his predecessor:

> "I salute my predecessor for his half-century of service to America, and thank the millions of men and women ...".

In this statement Clinton praises and thanks his predecessor and the people who have served their country. Due to the formal occasion of the ceremony, we may treat these acts as declarations. Common salutations, greetings and thankings would generally be classified as expressive acts.

The poetic function

The poetic function, which focuses on the message, selects elements from the code which draw attention to themselves and hence to the text. Poetic statements have been included in the speech to appeal to the hearers' emotions: "Though we march to the music of our time, our mission is timeless" (ll. 11-12). Such statements do not express much meaning if separated from the rest of the text.

The selection of "poetic elements" may be at the lexical level in terms of unusual collocations and metaphors, at the phonological level by means of rhyme- and rhythm-schemes and alliteration or at the syntactic levels by parallel structures, balanced constructions, interrupted movements, etc. (see *rhetorical features*).

The text act

The overall communicative function of Clinton's speech, i.e. the *text act*, is directive and commissive. It is directive because he wishes to appeal to the people to acknowledge their responsibility towards their country, and to take part in the work of improving the state of affairs in America. It is commissive because he, as President, wishes to commit the American people, as well as himself, to the task of changing the American society. The overall purpose of the speech is thus to make every American feel responsible towards, and important to, the developments in their own country. In a sense we could say that the many expressives and directives are put forward with the overall aim of committing the American people.

Clinton tries to win the trust and commitment of the American people. The method he uses is to appeal to their emotions as well as their pride in their country and their sense of responsibility towards their children and the future. By means of the frequent use of *we* he emphasizes that the task of improving American society rests on himself and the people together and he thus places himself on their side, making himself their friend and ally. In his effort to do so, he creates a strongly cohesive text with many lexical chains to bind his words and ideas together. The rhetorical devices he uses to achieve his aim is discussed in the following part.

Rhetorical features

An inaugural address is broadcast to millions of people all over the world. The amount of effort invested in the preparation of the communicative event probably surpasses that put into any other speech event. The address is premeditated at length, and rhetorical features of the speech are particularly salient.

In Clinton's address, there are several cohesive features which serve a rhetorical function. Obvious features are repetition and lexical chains at text level. These devices are used to emphasize important parts of the President's speech and to make his messages appear clear and appealing.

Repetition

The desire for change is emphasized through the many cases of repetition in the text including the term "change" both as a noun and as a verb (ll. 9, 32, 55). This change is for America and for the whole world as well; the word "world" is used many times (eg ll. 4, 5, 8, 18, 20, 31, 33, 57, 64, 71, 72). In connection with change many words occur which have something to do with changing the world into something better - into a new world: *renewal, reborn, renew, reinvent, new, remaking*. These words together with the term "change" help to convey Clinton's message of changing America/the world into a better place to live in.

Lexical chains

A lexical chain is a recurring pattern of related lexical items at text level. Lexical chains are an important cohesive device in Clinton's address. The dominant cohesive features of the inaugural concern the notion of change. This is of course far from incidental since Clinton's message to the American people is that America needs change. His repetition of *change* and the many items related to the notion of change form an entire cohesive chain of words related to the notion of change:

> l. 2 renewal; l. 5 reborn; l. 6 reinvent; l. 9 change, not change for change's sake; l. 10 change; l. 31 shaking and remaking; l. 32 change; l. 33 change; l.

55 dramatic change; l. 58 renewal; l. 60 a new season of American renewal; l. 61 renew.

It is clear that *renewal* and *change* are of vital importance to the President. Not only is the concept repeated in various forms in the text, but change is used in contrast to the the word *drift* (ll. 45-46), which has been used negatively. *Drifting* "has eroded our resource, fractured our economy, and shaken our confidence" and is the cause of the present bad state of American society.

The notion of change is introduced in the beginning of the speech and supported by reference to weather and the seasons. By contrasting "in the depth of the winter" with "spring", by referring to "the shadows of the Cold War" (l. 17) and "A world warmed by the sunshine of freedom" (l. 18), he implies that he wants to change the state of affairs from winter to spring and sunshine. He ends his cohesive chain with "a new season" (l. 60) and thereby confirms the optimism he introduces in the beginning of his speech. *Season* is used figuratively. Clinton represents the spring which is beginning and this is presented as opposed to the winter which is now over.

A way of expressing commitment to his task is the small cohesive chain of words which can be related to something religious and sacred:

l. 3 mystery; l. 3 ceremony; l. 8 Almighty; l. 10 life; l. 12 mission; l. 15 sacrifice; l. 26 ceremony; l. 66 sacrifice; l. 74 sacred responsibility.

Other cohesive chains in the text concern sacrifice and obligation:

l. 15 steadfastness and sacrifice; l. 66 it will require sacrifice;
l. 67 choosing sacrifice for its own sake;

l. 12 each generation must; l. 17 assumes new responsibilities; l. 44 we have to; l. 49 we must; l. 59 we pledge; l. 61 we must (twice); ll. 62, 64, 65, 67 we must.

In these cohesive chains there is a lot of repetition of lexical items. In particular in the final part of the text where the phrase *we must* is repeated seven times. This is done to emphasize the appeal to the people and to leave them with the impression that their new President is on their side.

By means of cohesive elements Clinton also manages to compare history to the present. Words and phrases are used first to describe what happened before, and then they are applied to statements concerning present problems:

l. 7 our founders boldly declared (past)

l. 61 we must be bold (present)

l. 9 not change for change's sake (past)

l. 67 sacrifice for its own sake + our own sake (present).

By using this technique, Clinton appeals to the pride of the people of the history and their country and the American ideals, and at the same time he suggests that the need for change is not at all new to America. By this suggestion he tries to convince people that they do not need to fear change; it has led to triumph before (over Depression, fascism, and Communism l. 16), and it will do so again.

The frame of American ideals includes:

> l. 5 democracy; l. 7 independence; l. 10 ideals; l. 10, life, liberty, pursuit of happiness; l. 18 responsibilities; l. 20 prosperity; l. 29 competition; l. 34 compete and win; l. 41 freedom; l. 47 challenges; l. 48 strengths; l. 56 democracy; l. 65 compete for every opportunity; l. 67 for our own sake; l. 68 family; l. 72 ideals; l. 74 sacred responsibility.

A frame of items referring to economy and business also involves important aspects:

> l. 20 economy; l. 21 business failures; l. 27 commerce; l. 29 earn; l. 29 competition; l. 34 compete; l. 34 win; l. 36 work; l. 36 cost of health care; l. 40 bankrupt;
> l. 40 enterprises; l. 46 economy; l. 57 engine; l. 63 invest; l. 64 cut massive debt; l. 65 compete.

Finally, a world frame is used to emphasize the importance of USA as a world leader (cf. the many repetitions of the word *world* listed above): l. 13 nation, l. 29 universal; l. 73 planet.

Contrasts are used frequently, for example threat/security, eastern/western, totalitarian/democracy, good/evil, Cold War/sunshine of freedom, etc.

The President's use of verbs which indicate decomposition: "eroded", "fractured" and "shaken" (ll. 46-47) emphasize the serious situation that must be changed - they form a contrast to the positive words referring to the change into a new and better world.

Metaphors and unusual collocations

Metaphors are representative of figurative speech. The essence of metaphor is understanding and experiencing one kind of thing in terms of another (Lakoff and Johnson 1980: 3). Features not strictly present are transferred to an object, a person, a state of affairs, etc.:

> l. 4 *We force the spring*; l. 5 *A spring reborn*; l. 6 *reinvent America;* l. 11 *we march to the music of our time*; l. 12 *each generation must define*; l. 15 *sacrifice triumphed*; ll. 17-18 *a generation raised in the shadows;* l. 18 *a world warmed by sunshine;* l. 24 *news travelled;* l. 32 *make change our friend*; l. 36 *the cost of health care threatens*; l. 53 *construct from these crises the pillars*; l. 56 *this is our time;* l. 57 *engine of our own renewal*; l. 60 *a new season of American renewal*; l. 70 *a child's eyes wander into sleep;* l. 73 *borrowed our planet.*

There are a few unusual collocations:

> l. 18 *the sunshine of freedom*; l. 28 *Technology is almost magical*; l. 29 *peaceful competition*; l. 74 *sacred responsibility.*

Metaphors and unusual collocations attract attention to the expression, thereby underscoring the significance of the message. At the same time, they make the text interesting to read or listen to.

Alliterations

Characteristic of the text is also the frequent use of alliterations:

> l. 11 <u>m</u>arch to the <u>m</u>usic; l. 22 <u>d</u>eep <u>d</u>evisions; l. 22 <u>i</u>ncreasing <u>i</u>nequality; l. 25 <u>s</u>ights and <u>s</u>ounds; l. 27 <u>c</u>ommunication and <u>c</u>ommerce; l. 30 <u>p</u>rofound and <u>p</u>owerful; l. 31 <u>sh</u>aking and <u>r</u>emaking; ll. 44-45 <u>s</u>trong <u>s</u>teps; l. 57 <u>e</u>nvy and <u>e</u>ngine; l. 60 <u>d</u>eadlock and <u>d</u>rift.

These alliterations help create a certain rythm in the text, and at the same time they add a poetic element to the text.

Parallel structures

Parallel structures, which occur within the sentence, may be 1) asyndetic (no coordinator), 2) monosyndetic (one coordinator), or 3) polosyndetic (several coordinators):

1) X, Y, Z
2) X, Y and Z
3) X and Y and Z.

The use of parallel structures is an extremely frequent device in Clinton's address. It is used to attract attention in a number of different syntactic constructions, as subject, object, apposition, and parallel clauses:

subject:

> There is no clear division today between what is foreign and what is domestic-*the world economy, the world environment, the world AIDS crisis, the world's arms race* affect us all
>
> *Our hopes, our hearts, and our hands* are with those on every continent who are building democracy and freedom
>
> But *no President, no Congress, no government* can undertake this mission alone

verb:

> Profound and powerful forces are *shaking and remaking* our world

predicate:

> and that drifting *has eroded our resource, fractured our economy, and shaken our confidence*

object:

> When our founders boldly declared *America's independence to the world and our purposes to the Almighty*, they knew America, to endure, would have to change
>
> Communism's collapse has called forth *old animosities* and *new dangers*

verb + object

> we know we have to *face hard truths and take strong steps*

apposition:

> change to preserve America's ideals - *life, liberty, the pursuit of happiness*

pre-modifiers:

> *the sights and sounds* of this ceremony
>
> *profound and powerful* forces are shaking and remaking our world
>
> *the vision and will* of those who came before us

post-modifiers:

> Today, a generation raised in the shadows of the Cold War assumes new responsibilities in a world *warmed by the sunshine of freedom but threatened still by ancient hatreds and new plagues*

prepositional phrases:

> *by the words we speak and the faces we show the world,* we force the spring

> triumphed *over depression, fascism, and Communism*

> Raised in unrivaled prosperity, we inherit an economy still the world's strongest, but weakened *by business failures, stagnant wages, increasing inequality, and deep divisions among our people.*

> Powerful people maneuver for position and worry endlessly *about*

> *who is in and who is out, who is up and who is down*

adverbials:

> news travelled slowly *across the land on horseback and across the oceans by boat*

adverbial clauses:

> But,
> *when most people are working harder for less,*
> *when others cannot work at all,*
> *when the cost of health care devastates millions and threatens to bankrupt many of our enterprises, great and small,*
> *when millions of poor children cannot even imagine the lives we are calling them to lead -*
> we have not made change our friend

> When our vital interests are challenged,
> or the will and conscience of the international community defied,
> we will act -
> *with peaceful diplommacy when possible,*
> *with force when necessary*

parallel clauses:

> Now,
> the sights and sounds of this ceremony are broadcast instantaneously to billions around the world.
> Communications and commerce are global;
> investments is mobil,
> technology is almost magical;
> and ambition for a better life is universal

Balanced sentences

A balanced sentence consists of two (or more) fairly balanced parts which are often contrasted:

> The Americal people have summoned the change we celebrate today.
> *You have raised your voices in an unnistakable chorus.*
> *You have cast your votes in historic numbers,*
> *and you have changed the face of the Congress, the presidency, and the political process itself.*
> *Yes, you have forced the spring.*
>
> Not change for change's sake,
> *but* change to preserve America's ideals
>
> Our democracy must be
> not only the envy of the world
> *but* the engine of our own renewal
>
> not choosing sacrifice *for its own sake,*
> but *for our own sake*
>
> It will not be easy;
> it will require sacrifice.
> *But* it can be done,
> and done fairly
>
> Posterity is the world to come -
> *the world to whom we hold our ideals,*
> *from whom we have borrowed our planet,*
> *and to whom we bear sacred responsibility*
> (balanced apposition)

> We must do what America does best;
> *offer opportunity to all*
> *and demand responsibility from all*
> (balanced parallel strcutures)
>
> It is time to break the bad habit of expecting something for nothing,
> *from our government or from each other*
>
> Let us take more responsibility,
> *not only for ourselves and our families*
> *but also for our communities and our country*
> (balanced parallel structures)

with substitution:

> Though our challenges are fearsome,
> so are our strengths.

Parallel structures and balanced constructions are also typical features of Kennedy's inaugural address, for example in his famous request: "Ask not what your country can do for you, ask what you can do for your country" and in his challenging statement, "Let every nation know, whether it wishes us well or ill, that we shall *pay any price, bear any burden, meet any hardship, support any friend, oppose any foe* to assure the survival and the success of liberty". Parallellisms are also present, although less successfully exploited, in President Bush's inaugural.

Iconicity

A typical cohesive feature of the text is iconic linkage. It is a repetitive pattern occurring at text level to attract attention and to bind parts of the text together.

> *To renew America, we must* be bold.
> *We must* do what no generation has had to do before.
> *We must* invest more in our own people and in our own future,
> and at the same time cut the massive debt.
> And *we must* do so in a world in which we must compete for every opportunity
>
> *To renew America*, we must revitilize our democracy
>
> And *we must* care for one another
>
> *To renew America*, we must meet challenges abroad as well as at home

> *Let us* resolve to reform our policits, so that
>
> *Let us* put aside personal advantage, so that
>
> *Let us* resolve to make our government a place for...
>
> *Let us* give this capital back to the people to whom it belongs
>
> *let us* begin with energy and hope, with faith and discipline,
> and *let us* work until our work is done.

Iconicity + parallel structures

> Today, we do more than celebrate;
> we rededicate ourselves to the very idea of America:
> *An idea* born in revolution and renewed through two centuries of challenge;
> *An idea* tempered by the knowledge that, for fate, we - the fortunate and the unfortunate - might have been each other;
> *An idea* ennobled by the faith that our nation can summon from its diversity the deepest measure of unity;
> *An idea* infused with the conviction that America's long heroic journey must go forever upward
>
> Posterity is *the world to come - the world for whom* we hold our ideals, *from whom* we have borrowed our planet, and *to whom* we have sacred responsibility.

Interrupted movement

Interrupted movements have a modifying element placed between subject and verb. This element often adds a personal tone to the text.

> This beautiful capital, *like every capital since the dawn of civilization*, is a place of intrigue and calculation.
>
> The brave Americans serving our nation in the Persian Gulf, *in Somalia, and whereever else they stand,* are testament to our resolve.

As a contrast to interrupted movement, which is a complex construction with separation of subject and verb, we find short concise sentences:

> Americans deserve better.

Sound bites

Inaugurals are supposed to contain *sound bites* to be remembered by. A sound bite is a 'bite-sized' chunck that the media, and in particular the audio-visual media, can make prominent in their snack-sized news broadcasts (WAUDAG 1990: 193). We all remember Kennedy's famous words:

> Ask not what your country can do for you, ask what you can do for your country.

In Clinton's address, the notion of change is probably the most prominent clue to the understanding of his message:

> Not change for change's sake,
> but change to preserve America's ideals -
> life, liberty, the pursuit of happiness.

Note also bites such as *To renew America* which occurs several times and short simple sentences, such as *My fellow citizens, this is our time. Let us embrace it.* Clinton emphasizes change: "We are to make change our friend, not our enemy", and he ends with "We have heard the trumpets. We have changed the guard. And now each in our own way, and with God's help, we must unswer the call".

The inaugural address as a genre

When establing the genre, what are the criteria? The rhetorical situation is obvious - it is a unique situation. How about the communicative purpose? In the beginning of my article I pointed out that inaugurals differ from most political speeches in that they are not constructed to make people vote for a certain party, person, or policy. What then is the communicative purpose? Some researchers have looked at inaugurals as a special genre with a special symbolic function.

The inaugural address of American presidents has been treated as a special genre involving a special symbolic function (e.g. Hart 1984, Campbell and Jamieson 1986, Gronbeck 1986). A recent study by WAUDAG (1990) sees the inaugural address as a restoration of ideological normality, which is to say invisibility, in American public discourse. The partisan strife of a presidential campaign heightens public awareness of the relation of discourse to power.

At the moment of inauguration, the President-elect, who has prevailed in the partisan struggle, becomes the President of all Americans, the head of the government and head of state, a figure who rises above divisions to re-cover the common perceptions, values and concerns of the nation.

Another function is to invoke the notion of the "Plural Presidency". An inaugural address is the first opportunity for the president to use the official voice. In the words of Hart (1984: 58) "The president speaks not for himself but for his people: he uses cadences reserved for majestic moments; he borrows his lexicon from his predecessors."

Plurality is further underscored by what we know of the drafting of presidential speeches: the president seems to function more as a selector than an originator of words, as well as a selector of his speech writers. So if we are to answer the question *Who speaks?* the answer must be: It is not the individual, not the situation, but the institution. The inauguration proves a call to action for others: for members of the government, for President Bush for the volunteerism of "a thousand points of lights", for Clinton "a call to service in the valley".

WAUDAG (1990: 189)) share the belief that "the analysis of inaugural addresses must be appropriately situated", but they find that the relevant situation lies "not so much in other inaugural addresses, or in inauguration as a symbolic act, but in the address as a restoration of "ideological normality", which is invisibility in American public discourse. In their analysis of President George Bush's Inaugural Speech of January 1989, they view his address as

> a restoration of ideological normality, which is to say invisibility, in American public discourse: the President acts to re-cover (in many senses) the common values and concerns of the 'the nation'. (1990: 189)

The address works to construct "both the privileged public figure of the President, and a spirit of collective identity and consent for the audience" in which the presidential role is described as that of "politically detached steward of a gendered status quo rather than agent of change" (1990: 189).

In this paper it is argued that although inaugural addresses share common features characteristic of the symbolic act of inauguration, specific features contributed by the particular president (and/or his advisers) may be found as well. In the case of President Bill Clinton's inaugural address, the outstanding characteristic is not that of "a politically detached steward in a gendered status quo"; on the contrary, his message to the nation is that the nation change, the nation must work together to renew America, "Not change for change's sake, but change to preserve America's ideals - life, liberty, the persuit of happiness". The President establishes himself as the leading figure in instigating the desired change (cf. the conclusion of the analysis of communicative functions in which directive/commissive acts are established as the text act).

Are then all inaugurals the same?

Not all presidents are eloquent to the same degree. Kennedy was praised for his elegant speeches. I find that Clinton is more eloquent than Bush, who has often been criticized for being vague to the extent of jarring colloquialism:

> To my friends — and yes, I do mean friends — in the loyal opposition — and yes, I do mean loyal opposition, I put out my hand, I am putting out my hand to you, Mr Speaker, I'm putting out my hand to you Mr Majority Leader, for this is the thing, this is the age of the offered hand. And we can't back clocks, and I don't want to.

Bush proposes and invites the identification of himself and the audience as neighbours and friends. Later he offers the hand of friendship to the 'loyal opposition'. Thus the illusion that Bush is indeed neighbour and friend is projected. However, the resulting quality of casual vagueness was promptly ridiculed in comic strips.

As pointed out, I also detect differences in attitude between Bush's and Clinton's addresses. Bush remains a detached observer of affairs. He is not a responsible agent promoting freedom in the world, rather change comes with the wind ("A new breze is blowing", and "totalitarian ideas are being blown away"). In contrast to Bush as the aloof and impassive watcher of the scene, my analysis has shown that Clinton passes judgement, and is willing to act as an agent of the changes he prophesizes.

The similarities between inaugural addresses are found in the move structure.

Move structure

Important moves are confirming the presidential role, which is done through the collaborative construction of the president as a (collective) public figure, and the acceptance of the official voice.

Constructing the presidential identity

The President must restore himself as the President of the American people. After the partisan strife, the President-elect must reestablish himself as the President; he is the head of government, the head of state, who rises above political divisions, a public figure all Americans can identify with and salute as their President. The audience is invited to participate in the construction of the presidential identity. Both Kennedy and Clinton open their speech with the plural *we*:

> We observe today not a victory of a party but a celebration of freedom
> My fellow citizens: Today, we celebrate the mystery of American renewal.

To achieve collective identity, President Bush addresses the audience as friends and neighbours and he even reaches out to the opposition:

> To my friends — and yes, I do mean friends — in the loyal opposition — and yes, I do means loyal: I put out my hand.

In their construction of presidential identity, all three presidents exploit the nation's patriotism, they lead the audience to reflect on salient points in history, on great figures in the past, and finally, with patriotic pride, to ascribe/connect the new President to this list of famous men. Stressing the continuity, George Bush finds that George Washington would "be gladdened by this day", i.e. by having George Bush as the latest descendant in the presidential line. For Bill Clinton "our founders boldly declared America's independence to the world and our purposes to the Almighty", he explicitly mentions George Washington and Franklin Roosevelt, and there are a number of allusions to President John F. Kennedy.

The official voice

The President must accept the official identity assigned to the president on Inauguration Day. In the modern inaugural, the president does not speak for himself, he speaks for his people. In doing so "he uses cadences reserved for majestic moments; he borrows his lexicon from his predecessors" (Hart 1984: 58, quoted in WAUDAG 1990: 193). This does not mean, however, that inaugurals are merely routine without ideology. Even though the new president borrows from his predecessors and from the tradition, he has chosen which people to identy with, and consequently which cadences and words to speak. From the present analysis, it is apparent that President Clinton's speech is widely different from President Bush's, but shows clear parallels to that of President John F. Kennedy.

Other typical moves

Typical moves also include addressing the audience, thanking the predecessor, reference to the ceremony, patriotism and liberty, pride in American history and glory, religion, family values and tradition.

President Bush thanks President Reagan, a man "who has earned a lasting place in our hearts - - and in our history" on behalf of the nation for "the wonderful things" he has done for America, while President Bill Clinton salutes "my predecessor for his half century of service to America" and thanks "the millions of men and women whose steadfastness and sacrifice triumphed over depression, facism, and communism".

George Bush has "just repeated word-for-word the oath taken by George Wahington 200 years ago" placing his hand on "the Bible on which he placed his". President John F. Kennedy swore "before you and Almighty God the same solemn oath our forebears prescribed nearly a century and three quarters ago", while President Clinton embeds this reference in his report on changes in The United States: "When George Washington first took the oath I have just sworn to uphold..."

The aspect of restoring himself as the President of the American nation, which is an important issue in Bush's speech, is only briefly touched upon in President John F. Kennedy's speech: "We observe today *not a victory of a party* but a celebration of freedom, symbolizing an end as well as a beginning, signifying renewal as well as change", while in President Bill Clinton's speech, this aspect is only indirectly present in his collective address form *we*.

Religious overtones are prominent in President John F. Kenendy's speech. He takes the oath before Almighty God and calls on God to help him in his service. President Bush's first act as President is a prayer, and he ends his speech in the same way: "God bless you. God bless the United States of America". For Clinton, his presidency is a *mission, a call for service in the valley* for him and the American people and he ends his address by invoking God's help: *We have heard the trumpets. We have changed the guard. And now each in our own way, and with God's help - we must answer the call.*

Resemblance to other political speeches

The inaugural certainly stands out due to the uniqueness of the situation, which involves taking the oath, constructing the presidential identity and accepting the official voice. At the same time, the inaugural address shares a number of features with most other political speeches, in particular the aspects mentioned under communicative functions and rhetorical strategies. A number of aspects are highlighted in the following:
- the involvement of the audience (complex participation)
- the direct address/the appeal to work together
- use of familiar symbols and values, a sense of communion
- attempted manipulation

- stylistic features of political rhetoric (e.g. repetition, use of metaphor, parallel structures, balanced constructions, iconic linkage, lexical chains, etc.
- argumentative text type.

As most political speeches, the inaugural address is representative of the argumentative (persuasive) text type.

Kennedy Inauguration Address
Bush Inauguration Address
Clinton Inauguration Address

Notes

1 Austin orginally introduced five categories: Verdictives, exercitives, commissives, behabitives, and expositives. For a reclassification of these categories into representatives, directives, commissives, expressives, and declarations, see Searle (1976).
2 For a classification of *assessives*, see Vestergaard, this volume.
3 Performative verbs explicitly express the illocutionary force of an utterance.
4 Declarations are typical in legal language when laying down the law.
5 WAUDAG stands for the University of Washington Discourse Analysis Group: George L. Dilon, Anne Doyle, Carol M. Eastman, Susan Kline, Sandra Silberstein and Michael Toolan.

References

Austin, J.L. 1962. *How to do Things with Words*. New York: Oxford University Press.
Bhatia, V.K. 1993. *Analysing Genre: Language Use in Professional Settings*. London: Longman.
Campbell, K.K. and Jamieson, K.H. 1986. "Inaugurating the presidency". In H.W. Simons and A.A. Agazarian (eds), *Form, Genre, and the Study of Political Discourse*. University of South Carolina Press, 203-25.
Gronbeck, B.E. 1986. "Ronald Reagan's enactment of the presidency in his 1981 inaugural address". In Simons and Agazarian, pp. 226-45.
Hart, R.P. 1984. *Verbal Style and the Presidency*. Orlando: Academic Press.
Hatim, B. and Mason, I. 1990. *Discourse and the Translator*. New York: Longman.
Jakobson, R. 1960. "Linguistics and poetics". In Sebeok, *Style in Language*. M.I.T. Press.
Lakoff, G. and Johnson, M. 1978. *Metaphors We Live By*. Chicago: University of Chicago Press.
Searle, J.R. 1976. "The classification of illocutionary acts". *Language in Society* 5: 1-24.

Swales, J.M. 1990. *Genre Analysis. English in Academic and Research Settings.* Cambridge: Cambridge University Press.

Traugott, E.C. and Pratt, M.L. 1980. *Linguistics for Students of Literature.* New York: Harcourt Brace Jovanovitch.

WAUDAG[5] 1990. "The rhetorical construction of a President". *Discourse & Society* vol. 1(2): 189-200.

The text of President Clinton's inaugural address (1993):

My fellow citizens: Today, we celebrate the mystery of American renewal. This ceremony is held in the depth of winter. But, by the words we speak and the faces we show the world, we force
5 the spring. A spring reborn in the world's oldest democracy, that brings forth the vision and courage to reinvent America. When our founders boldly declared America's independence to the world and our purposes to the Almighty, they knew America, to endure, would have to change. Not change for change's sake, but
10 change to preserve America's ideals - life, liberty, the pursuit of happiness. Though we march to the music of our time, our mission is timeless. Each generation must define what it means to be American. On behalf of our nation, I salute my predecessor for his half-century of service to America, and thank the millions
15 of men and women whose steadfastness and sacrifice triumphed over Depression, fascism, and Communism. Today, a generation raised in the shadows of the Cold War assumes new responsibilities in a world warmed by the sunshine of freedom but threatened still by ancient hatreds and new plagues. Raised
20 in unrivaled prosperity, we inherit an economy still the world's strongest, but weakened by business failures, stagnant wages, increasing inequality, and deep divisions among our people. When George Washington first took the oath I have just sworn to uphold, news traveled slowly across the land on horseback and
25 across the oceans by boat. Now, the sights and sounds of this ceremony are broadcast instantaneously to billions around the world. Communications and commerce are global; investment is mobile; technology is almost magical; and ambition for a better life is universal. We earn our livelihood in peaceful competition
30 with people all across the earth. Profound and powerful forces are shaking and remaking our world, and the urgent question of our age is whether we can make change our friend and not our enemy. This new world has already enriched the lives of millions of Americans who are able to compete and win in it. But when
35 most people are working harder for less; when others cannot work at all; when the cost of health care devastates millions and

threatens to bankrupt many of our enterprises, great and small; when fear of crime robs law-abiding citizens of their freedom; and when millions of poor children cannot even imagine the lives we are calling them to lead - we have not made change our friend. We know we have to face hard truths and take strong steps. But we have not done so. Instead, we have drifted, and that drifting has eroded our resource, fractured our economy, and shaken our confidence. Though our challenges are fearsome, so are our strengths. Americans have ever been a restless, questing, hopeful people. We must bring to our task today the vision and will of those who came before us. From our revolution to the Civil War, to the Great Depression to the civil rights movement, our people have mustered the determination to construct from these crises the pillars of our history. Thomas Jefferson believed that to preserve the very foundations of our nation, we would need dramatic change from time to time. My fellow citizens, this is our time. Let us embrace it. Our democracy must be not only the envy of the world but the engine of our own renewal. There is nothing wrong with America that cannot be cured by what is right with America. So today, we pledge that the era of deadlock and drift is over - a new season of American renewal has begun. To renew America, we must be bold. We must do what no generation has had to do before. We must invest more in our own people and in our own future, and at the same time cut our massive debt. And we must do so in a world in which we must compete for every opportunity. It will not be easy; it will require sacrifice. But it can be done, and done fairly, not choosing sacrifice for its own sake, but for our own sake. We must provide for our nation the way a family provides for its children. Our Founders saw themselves in the light of posterity. We can do no less. Anyone who has ever watched a child's eyes wander into sleep knows what posterity is. Posterity is the world to come - the world for whom we hold our ideals, from whom we have borrowed our planet, and to whom we bear sacred responsibility.

Genres in Conflict

Genres in Conflict[1]

VIJAY K. BHATIA
City University of Hong Kong

Introductory

Genres are dynamic constructs, even though they are essentially seen as embedded in conventions associated with typical instances of language use in social, academic or professional settings. In whatever manner one may identify them, whether as *a typification of rhetorical action*, (Miller 1984, and more recently, Berkenkotter and Huckin 1995), as *staged, goal oriented social processes*, (Martin 1993), or as *shared communicative purposes*, (Swales 1990, Bhatia 1993), an understanding or a prior knowledge of conventions is considered essential for its identification, construction, interpretation, use and ultimate exploitation by members of specific professional communities to achieve socially recognised goals with some degree of pragmatic success. This may give a somewhat misleading impression that generic forms are always standardised and static. However, more recently, it has also been pointed out that generic forms are rather dynamic in a number of ways. Starting with genre as typification of rhetorical action, it will be wrong to assume that rhetorical situations always recur exactly the same way and hence will require an exactly similar response, though there may be certain aspects of rhetorical situation which may typically recur. Members of a specific professional community often use their generic knowledge to recognise 'relevant similarities' (Miller 1984) in order to respond to such recurring or even novel rhetorical situations to create meanings in forms that are shared by other members in that community. That probably makes Berkenkotter and Huckin (1995: 6) point out that although genres are associated with typical socio-rhetorical situations, they "are inherently dynamic rhetorical structures that can be manipulated according to conditions of use..." They rightly claim that

> Genres ... are always sites of contention between stability and change. They are inherently dynamic, constantly (if gradually) changing over time in response to sociocognitive needs of individual users.

So, whenever a person responds to a somewhat changing socio-cognitive need, he may be required to adjust his response in the light of his experience of recognisable socio-rhetorical conventions, which may make it necessary for him to generate slightly varied or even new generic form(s).

If, on the other hand, we take *shared communicative purpose* as a privileged criterion to identify generic constructs, one may still find it necessary to account for variation in generic forms. Although genres are based on conventionalised, institutionalised, and to a large extent, standardised linguistic behaviour in various professional and academic settings, they have propensity for mixing and embedding. This gives considerable tactical freedom to expert members of the discourse community in question to manipulate generic resources and conventions in order to express either a set of communicative purposes, rather than a single communicative purpose, or simply to express 'private intentions' within the framework of 'socially recognised communicative goal' (Bhatia 1995).

Although it is true that most of the rhetorically situated texts have their 'generic integrity' (Bhatia 1993, 1994) as a result of a high degree of conventionalisation, it is still possible and often is the case that the expert members of the specialist community manage to exploit these conventions by mixing two or more established forms to create new hybrid forms. However, such innovative exploitations are invariably realised within rather than outside the broad generic boundaries, whichever way one may draw them, in terms of recurrence of rhetorical situations (Miller 1984), consistency of communicative purposes (Swales 1990), regularity of stages in goal-directed activities (Martin 1993) or a combination of these (Bhatia 1993). The freedom to innovate has its own constraints, in the sense that, if one were to flout generic conventions, it could potentially risk miscommunication or be noticed as odd by the established members of a professional community.

The dynamic complexity of academic and professional communication is further increased by the role of multi-media, explosion of information technology, multidisciplinary contexts of the world of work, increasingly competitive professional environment, and above all, the overwhelmingly compulsive nature of promotional and advertising activities, so much so that our present-day world of work is being increasingly identified as a "consumer culture" (Featherstone 1991). The inevitable result of this is that many of the institutionalised genres, whether they are social, professional or academic, are seen as incorporating elements of promotion. Fairclough (1993: 141), referring to such changes in discursive practices, points out,

> ...there is an extensive restructuring of boundaries between orders of discourse and between discursive practices; for example, the genre of consumer advertising has been colonizing professional and public service orders of discourse on a massive scale, generating many new hybrid partly promotional genres...

As an instance of such a hybrid genre, Fairclough (1993) discusses the case of contemporary university prospectuses, where, he highlights an increasing tendency towards marketization of the discursive practices of British universities. Bhatia (1995, 1997), in his discussion of genre mixing in professional discourse, gives examples from several settings, where genre-mixing and embedding is becoming increasingly common. He also mentions several instances where one may find an increasing use of promotional strategies in genres that are traditionally considered non-promotional in intent. The examples include job advertisements and academic introductions, where one can find quite explicit attempts to incorporate promotional elements in areas of academic and professional discourse, which traditionally have been predominantly informational. An interesting point to note in all these instances of genre-mixing and genre-embedding is that the two genres are somewhat compatible with each other, in that they do not show any conflict in their communicative purposes. A closer look at these instances will indicate that in most of them informative functions are colonised by promotional functions. As Bhatia (1993) points out the most popular promotional strategy in advertising has been *to describe a product or service in a positive manner*, which is seen as the information-giving function of language. These two functions of language, i.e., informational and promotional are therefore unlikely to create tension, if not entirely complementary. Although it may appear that this kind of genre-mixing is more common in genres that are less likely to create functional tensions, it will be somewhat premature to assume that it will always be the case. The purpose of this paper is to look at some of the instances of public discourse in Hong Kong which appear to give rise to conflicts in genre.

Legal discourse in Hong Kong

Public legal discourse in Hong Kong, especially close to the period of changeover of sovereignty from Britain to the People's Republic of China presents an interesting opportunity to study how political decolonisation can gradually create conditions for an equally interesting colonisation of some public discourse genres which have traditionally been regarded as conflicting, especially the promotional and legislative genres in the emerging political context. On the face of it, there seems to be hardly anything common between the two. Legislation, which is meant to control public life, has primarily a regulative communicative purpose. As Bhatia (1993) points out, legislative writing is used to impose obligations and to confer rights. However, in order to control the capacity of humans to wriggle out of their obligations and to stretch their rights to unexpected limits, legislation is generally intended to be precise, clear, unambiguous, on the one hand, and all-inclusive, on the other, although the two sets of intentions may appear to be somewhat contradictory in certain ways.

The other aspect of public discourse which had assumed increasing significance in the years leading to the changeover of sovereignty was the concern on the part of both, the British Government and that of the People's Republic of China (PRC), to assure the people of Hong Kong and also the rest of the world that transition of sovereignty was going to be smooth and free of any conflict. This was very important for both the governments in order to maintain confidence in the viability of Hong Kong as an important centre for international trade and finance. Although Hong Kong had been a British colony for almost 150 years, there was no doubt that Britain had made a very significant contribution to the making of what Hong Kong represents today, especially as a remarkable economic miracle in Asia. It was important for Britain to be seen as handing over Hong Kong gracefully and, at the same time, and perhaps more importantly, making sure that the confidence of the people of Hong Kong was in no way undermined. It could have become quite traumatic for many of the Hong Kong residents to adjust to communism after a consistent dosage of capitalism for such a long period of time. Besides, no government can afford to underestimate the diversificational interests of the industry in the age of rapidly expanding multinationalism.

People's Republic of China, on the other hand, would not have liked to rock the boat at that critical juncture. It was certainly not in the interest of the PRC government to send wrong signals to the residents of Hong Kong, and more importantly, to the people of Macau and Taiwan as well. Macau will be the next territory to be handed over to PRC in 1999 and, PRC Government has always claimed Taiwan to be part of the Peoples Republic of China and, efforts for the re-unification at times appear to assume increasing importance. The other factor that supports this hypothesis further is the economic freedom that PRC has given to its southern provinces, especially Shanzen, Guandong and Shanghai. One country, two systems, that was, and still is, the promise. Once again, it was not in the best interests of the PRC government to encourage conditions which might have been detrimental to the survival of Hong Kong as an economic miracle, certainly not during and immediately after the take over of the territory.

In the circumstances it is absolutely crucial for both the parties to be seen to be working towards a smooth and conflict-free transfer of sovereignty. The most important indication of their good intention was what is known as the Sino-British Joint Declaration signed by the two governments in 1984. It is an interesting document which, on the one hand, was intended to signal to the people of Hong Kong and, of course, to the rest of the world, the importance the two countries attached to smoothness of transition of power, while on the other, it was also meant to be the basis in the light of which all future disputes were to be considered and perhaps settled. As one can see, the Joint Declaration was intended to have two somewhat conflicting intentions, one *legislative*, to provide solutions to all future disputes, and the other *diplomatic*, to promote and to give expression to mutual understanding, and perhaps to postpone, or even avoid, if necessary, painful and difficult decisions on contentious and unresolved issues of potential conflict to be managed through further negotiations as and when desirable. The legislative function often requires clarity, precision, unambiguity and comprehensiveness in the expression of such issues, whereas the most characteristic feature of diplomatic function of language is its vagueness, indirectness, generality and flexibility of expression in order to avoid clarity or commitment to any specific interpretation of issues. Considered in this manner, the two communicative functions are conflicting and contradictory, to say the least. However, it is interesting to find both of them present in the same document.

Perhaps this duality of somewhat conflicting intentions is a typical feature of what in diplomatic contexts is also known as *memorandum of understanding*. If one were to look at the diplomatic intentions in such documents, they are rightly called memorandum of understanding. However, if one were to take seriously the other aspect of many of these documents, they can be just the opposite of what was intended in the first place and hence no better than *memorandum of (mis)understanding*.

The Sino-British joint declaration

The very first introductory section of the documents sets the tone of the document when it expressly states:

> The Government of the United Kingdom of Great Britain and Northern Ireland and the Government of the People's Republic of China have ... agreed that a proper negotiated settlement of the question of Hong Kong ... is conducive to the maintenance of the prosperity and stability of Hong Kong and to

the further strengthening and development of the relations between the two countries...

[Joint Declaration, 1984: 11]

In nutshell, the Joint Declaration is a political or more appropriately a diplomatic statement with the effect of a legislative intention. It worked extremely well as a political statement, but was pragmatically rather contentious as an instrument of legislative expression.

Let us look at some of the sections in more detail. Section 3(3) assigns all kinds of executive, legislative and judicial powers to the people of Hong Kong, with an assurance that the laws of Hong Kong will remain unchanged for 50 years after the transfer of sovereignty.

> The Hong Kong Special Administrative Region will be vested with executive, legislative and independent judicial power, including that of final adjudication. The laws currently in force in Hong Kong will remain basically unchanged.
>
> [Section 3 (3) of the Joint Declaration, 1984: 11]

However, the last subsection of section 3 also provides that these basic policies will be stipulated in a Basic Law of the Hong Kong SAR, which will be drafted by the National People's Congress of the PRC.

> The above-stated basic policies of the People's Republic of China regarding Hong Kong and the elaboration of them in Annex I to this Joint Declaration will be stipulated, in a Basic Law of the Hong Kong Special Administrative Region of the People's Republic of China, by the National People's Congress of the People's Republic of China, and they will remain unchanged for 50 years.
>
> [Section 3(12) of the Joint Declaration, 1984: 13]

Although Annex I is a little more elaborate than the main document, it is still far from the traditional legislative rigour of the British and, indeed the Commonwealth writing.

> After the establishment of the Hong Kong Special Administrative Region, the laws previously in force in Hong Kong (i.e. the common law, rules of equity, ordinances, subordinate legislation and customary law) shall be maintained, save for any that contravene the Basic Law and subject to any amendment by the Hong Kong Special Administrative Region legislature.
>
> The legislative power of the Hong Kong Special Administrative Region shall be vested in the legislature of the Hong Kong Special Administrative Region. The legislature may on its own authority enact laws in accordance with the

provisions of the Basic Law and legal procedures, and report them to the Standing Committee of the National People's Congress for the record. Laws enacted by the legislature, which are in accordance with the Basic Law and legal procedures shall be regarded as valid.

The laws of the Hong Kong Special Administrative region shall be the Basic Law, and the laws previously in force in Hong Kong and laws enacted by the Hong Kong Special Administrative Region legislature as above.

[Annex I, section II of the Joint Declaration, 1984: 15]

From these sections, it is clear that the maintenance of the so-called status quo in Hong Kong for the next 50 years after the transfer of sovereignty, which includes the executive, legislative and judicial systems, crucially depends on the contents of the Basic Law, which was not written at the time of the Joint Declaration in 1984. Obviously, at the time when Joint Declaration was signed, there was some understanding in terms of broad principles rather than on matters of detail. For the People's Republic of China, it was perfectly normal and simply a matter of routine, since much of their judicial system works on a system which is rather different from that which underlies much of Common Law. Civil Law, especially that used in the PRC, relies heavily on broad principles rather than expressions of detail. The system allows wide ranging flexibility to judges to construe, interpret and use broad legal principles to arrive at judgements. However, in the Commonwealth legal system, the court decisions are based on precedents and the laws are much more detailed and hence transparent. In legal drafting in the common law countries, especially in the United Kingdom, there is every attempt by the legal draftsman "to box the judge in the corner" (Bhatia 1982), whereas in the PRC judiciary is validly regarded as part of the executive, thus enjoying much greater powers of interpretation. Let me give some substance to this by considering examples from the two systems.

The Basic Law

The Basic Law, as drafted by the People's Republic of China on the 4th of April 1990, is once again an interesting case in sight. It is written in what comes very close to ordinary English, with very little elaboration on a number of crucial issues. It is, in fact, nothing more than expression of a set of basic guidelines, which is meant to govern the people of Hong Kong after the transfer of sovereignty. One of the important issues raised there is that of the status of the laws already in force in Hong Kong. Article 8 of the Basic Law states,

> The laws previously in force in Hong Kong, that is, the common law, rules of equity, ordinances, subordinate legislation and customary law shall be maintained, except for any that contravene this law, and subject to any amendment by the legislature of the Hong Kong special Administrative Region.

As one may see, like most other articles of the Basic Law, this one too is expressed in terms of a universally applicable principle, which is meant to cover every possible contingency one can think of in the context of pre-existing legislative machinery. One may be tempted to point out that there should be no serious problem in expressing legislation in terms of general broad-based principles. After all there are many civil law countries, where legal systems adopt such a strategy. The French legal system is a good example of this (See Bhatia, 1993 for illustration). However, a lack of elaboration and legal specification in French and perhaps other European legislative documents, is an important part of their legal systems, which is meant to allow flexibility and power to the judiciary in these countries. In the Common Law countries, on the other hand, as Sir Renton (1975: 36) points out, a range of qualifications provides a degree of legal specification which goes a long way to adequately safeguard 'the liberty, the purse, or the comfort of individuals' in a democratic society. As Caldwell (quoted in Bhatia 1982: 51) points out,

> ...if you extract the bare bones...what you end up with is a proposition which is so untrue because the qualifications actually negative it all...it's so far from the truth...it's like saying that all red-headed people are to be executed on Monday, but when you actually read all the qualifications, you find that only one per cent of them are...

There are a number of countries where laws and statutes are expressed in general terms, however, in the case of the Basic Law, there have been several clauses which have given rise to contesting interpretations, primarily because a number of issues were left unresolved and to make it worse, all this happened in the context of two very different legal systems interacting with each other, one with the most elaborate and exhaustive legislative style based on the model used in the U. K. and the other extremely diluted plain English expressions of principles captured in the Basic Law. Every time a new ordinance is considered or promulgated in Hong Kong, it becomes a matter of fresh negotiation.

The main difference between the two ways of drafting (the PRC and the British system) is that in the commonwealth drafting every effort is made to make the provision all-inclusive, in an effort to make it transparent for the reader, on the one hand, and to ensure that court cases are judged in a more or less similar manner, if the legally material facts are similar. The power is thus transferred from the judiciary to the legislative body, which may be the parliament, senate or any other. In the Civil Law countries, the style of drafting gives the judiciary much greater freedom to construe, interpret and apply legal rules. The two systems, although different in the way they are used, are justifiable in their own right. Nevertheless, they are two different systems, based on different ideologies and serve different purposes. In one case they are seen as transparent and almost direct instruments of controlling the behaviour of people, whereas in the other, they are seen as expressions of principles meant for guiding, rather than directly controlling the behaviour of people. Both are seen as embedded in different socio-political ideologies. However, when they come in contact with each other, it is only natural that people will construe and interpret them differently, causing tension wherever they find any room for a varying interpretation. In the following section from the Basic Law, for instance, many of the complex rights are expressed in terms of general principles, which may appear to be least contentious on their own, however, in the context of any specific configuration of facts, many of them can become extremely contentious, especially when new laws are drafted and promulgated by the legislature, as it did actually happen on many occasions immediately after the transfer of sovereignty on the 1st of July 1997. Ghai (1997), a prominent legal specialist on constitutional matters related to the Basic Law, identifies two major issues in the drafting of the Basic Law. The first one is the result of basic principles becoming self-executing in some contexts.

> The Basic Law has nine chapters. In accordance with the Chinese practice, the first chapter sets out the 'General Principles' which underlie the Basic Law. The genesis of the General Principles is China's 12 points presented during the Sino-British negotiations on the transfer of sovereignty and the 'basic policies' in the main text of the Joint Declaration...The General Principles are a prelude to a more detailed provisions, but sometimes they are 'self-executing' as with art. 7 which vests the management of land and natural resources in the HKSAR or art. 9 which makes English an additional official language.
>
> [Ghai, 1997: 65]

The other one, as he points out, is the consequence of lack of specificity in drafting.

> The two broad areas on which there was considerable contention were the relations between the Central Authorities and the HKSAR and the political structure of the HKSAR. China had fought off the British during the negotiations for the Joint Declaration on these issues, and an appearance of consensus was purchased at the expense of ambiguity and obfuscation.
>
> [Ghai, 1997: 61]

To illustrate the point, let us take up the articles 12 and 13, which legislate the relationship between HKSAR and the Central Authorities of the PRC.

> Article 12
>
> The Hong Kong Special Administrative Region shall be a local administrative region of the People's Republic of China, which shall enjoy a high degree of autonomy and come directly under the Central People's Government.
>
> Article 13
>
> The Central People's Government shall be responsible for foreign affairs relating to the Hong Kong Special Administrative Region.
>
> The Ministry of Foreign Affairs of the People's Republic of China shall establish an office in Hong Kong to deal with foreign affairs.
>
> The Central People's Government authorizes the Hong Kong Special Administrative Region to conduct relevant external affairs on their own in accordance with this Law.

As one can see, article 12, on the one hand gives "a high degree of autonomy" to the HKSAR region, while at the same time, quite categorically, puts it directly under the Central People's Government, without any further specification of any kind. In the absence of any further specification, the expression 'a high degree of autonomy' can only be interpreted on a case by case basis and hence becomes a matter of fresh negotiation, every time it is invoked in a particular context. An important characteristic of this style of drafting is its lack of transparency, resulting from its lack of specification and all-inclusiveness, which is exactly the opposite of what has been almost taken for granted in the Commonwealth Legal system. So, the real problem is not that it lacks adequate degree of specification and transparency, but that it is likely to be interpreted in the context of a vary different legal system, which regards elaborate specification as almost sacrosanct.

Discussion

In the preceding sections, I have made an attempt to identify two kinds of generic conflict: the first resulting from genre-mixing, and the second, resulting from a mixing of two very different rhetorical contexts in which the same genre is meant to be construed, interpreted, used and possibly exploited, which provides for an interaction of two very different legal systems. However, the resulting legal documents in both the cases seem to share at least one typical linguistic characteristic, namely, a lack of all-inclusive specification that is so very typical of legislative statements in the Commonwealth countries.

The Sino-British Joint Declaration is a classic example of genre-mixing, which combines two very different sets of communicative purposes, that of *regulating* behaviour of two governments, and the other that of *promoting* goodwill and understanding. In the first sense, it has the force of an international treaty, whereas in the second sense, it has the force of successful diplomacy. It is rare to find a mixed genre incorporating two very different sets of communicative purposes, except in specific forms of diplomatic communiqués. In a majority of mixed genres, academic introductions, advertisements, newspaper and other reports, for example, (see Bhatia 1995, 1997, Fairclough 1993) we rarely see a potentially conflicting sets of communicative purposes, especially of the kind we see in the case of joint declarations or memoranda of understanding. It is typical of diplomacy to make statements forcefully and yet get away from any commitments that may follow from those declarations.

The other issue, which arises as a result of the same genre being interpreted in the context of two very different legal contexts, the Civil Law system used in People's Republic of China and the Common Law system, inherited from the British Commonwealth used in Hong Kong is more subtle than the straightforward mixing of two conflicting communicative purposes. In the case of the Basic Law, it is a single genre, with its own integrity embedded in the legal system of the People's Republic of China, and yet when it is used in two different rhetorical contexts, conflicts arise once again.

An important point that is bound to arise from such a situation is the question of interpretation of the Basic Law, which may have one interpretation if construed in Hong Kong Special Administrative Region (HKSAR) and a different one if interpreted in the mainland. This may be the result of the several different factors, some which include, somewhat different assumptions underlying the two legal systems, the way the legislative genre is constructed in the two systems, the nature and extent of power and freedom of interpretation vested in the judiciary in the two legal systems, and above all, the mutual misconceptions about the conflicting political ideologies in the two rhetorical situations. In this context, it is interesting to note that the Special Group on Law in Hong Kong, reporting on the powers of interpretation and amendment of the Basic Law also highlights the potential areas of conflict resulting from a lack of adequate specification. Let us look at the provision for the amendment of the Basic Law under the Joint Declaration first.

Amendment of the Basic Law

Provision under the Joint Declaration:

The National People's Congress of the People's Republic of China shall enact and promulgate a Basic Law of the Hong Kong Special Administrative Region of the People's Republic of China (hereinafter referred to as the Basic Law) in accordance with the Constitution of the People's Republic of China...

(Para 1, Section 1, Annex 1)

Although the Joint Declaration clearly indicates that the National People's Congress will enact the Basic Law, it does not provide any indication for its amendment or interpretation, which, like many other contentious issues was left open to be resolved by future negotiations. This lack of specification and all-inclusiveness led to the setting up of a special group on the Basic Law in Hong Kong, which considered the issue of amendment of the Basic Law in the context of the Joint Declaration and the Chinese Constitution, under which the Basic Law was to be enacted and promulgated, and suggested that there were several possible interpretations, most of them arising from the fact that this issue had failed to find adequate specification in any of the existing provisions, either in the Joint Declaration or in the Chinese Constitution.

Provisions under the Chinese Constitution:

Article 31
The state may establish special administrative regions when necessary. The systems to be instituted in special administrative regions shall be prescribed by law enacted by the National People's Congress in the light of the specific conditions.

Article 62
The National People's Congress exercises the following functions and powers:

(3) to enact and amend basic statutes concerning criminal offences, civil affairs, the state organs and other matters;

(12) to approve the establishment of provinces, autonomous regions, and municipalities directly under the Central Government;

(13) to decide in the establishment of special administrative regions and the systems to be instituted there:

Article 64

Amendments to the Constitution are to be proposed by the Standing Committee of the National People's Congress or by more than one-fifth of the deputies to the national People's Congress and adopted by a majority vote of more than two-thirds of all the deputies to the Congress.

Statutes and resolutions are adopted by the majority vote of more than one-half of all the deputies to the national People's Congress.

The Special Group, thus raise the following pertinent points on the Relationship between the Central Government and the HKSAR.

It is generally held that since the Basic Law shall be enacted by the NPC, it and only it shall have the power to amend the Basic Law.

Who can propose amendments to the Basic Law?

Opinion A

Members note that under the Chinese Constitution, the NPC and the State Council have the right to propose amendments to basic statutes. The by-law of the NPC provides that 30 members of the NPC together can initiate the proposal to amend the basic statutes. Therefore the NPC should have the power to initiate amendment to the Basic Law.

Opinion B

Nevertheless, it was proposed that the SAR Government or the SAR legislature should have the right to initiate proposals for amendments to the Basic Law.

Opinion C

An opposing view to this proposal is that if people of foreign nationals are permitted to sit on the SAR legislature, then they may change a very important law of China.

Opinion D

The Hong Kong delegates to the NPC shall have the sole right to propose amendments to the Basic Law.

Opinion E

Directly elected representatives from Hong Kong shall have the sole right to propose amendments to the Basic Law.

Opinion F

The Hong Kong Legislature shall have the sole right to propose amendments to the Basic Law with no restriction on the composition of the Legislature provided they are all local inhabitants as stated under the Joint Declaration.

It is interesting to note that the most important issue of a multiplicity of interpretations arises from a general feeling that "since the Basic Law shall be enacted by the NPC, it and only it shall have the power to amend the Basic Law". A number of misconceptions of this kind arise as a result of the style of legislative writing the two legal systems have conventionally adopted. Legislative style often associated with the mainland legal system is considered less detailed, and therefore less transparent, thus giving extensive interpretative power to the judiciary, which is often seen as less independent in the PRC. In Hong Kong, on the other hand, legislative style is detailed and all-inclusive in the true Commonwealth tradition, always making an effort "to box the judge firmly into a corner" (Caldwell, quoted in Bhatia 1982), thereby giving supreme authority to the legislative body elected by the people rather than the government appointed judiciary, which is consistent with the expectations in a typical democratic form of government. Much of the variation and potential conflict in the interpretation of the same constitutional document therefore, is the result of a possible interaction of the two conflicting rhetorical contexts in which the document is likely to be construed, interpreted and eventually used and the events of the past few months, especially the controversies that have arisen during this period, confirm this point of view.

To conclude, the paper investigates a complex, dynamic and somewhat chaotic area of public discourse in order to see if there is any motive or underlying pattern in the scheme of rather conflicting textual details, which give this particular genre its established status. It is interesting to see that although genre-mixing is often seen in genres which display a natural compatibility in their communicative intentions, it is by no means always the case. There can be and certainly are well-established instances of mixed genres, which typically combine two very different communicative purposes within a single generic artefact. This tendency to combine two or more seemingly incompatible communicative goals within a single generic construct further demonstrates the versatility of the generic framework, on the one hand, and the human capacity to exploit generic conventions to create new hybrid forms of discourse to meet the challenges of novel and rapidly changing rhetorical situations, on the other. The universe of real life communicative behaviour is complex, dynamic, largely unpredictable and full of surprises and the field of discourse and genre analysis is relatively young. One may only need to look around oneself for interesting and innovative marvels of human communicative creations embedded in novel and not so novel rhetorical contexts to realise how far one has to go.

Notes

1 This is the revised version of the paper presented at the AAAL Conference held in Seattle, Washington, USA, March (14-17) 1998.

References

Berkenkotter, C., and Thomas N. Huckin 1995. *Genre Knowledge in Disciplinary Communication — Cognition/Culture/Power.* New Jersey. Lawrence Erlbaum Associates, Publishers.

Bhatia, V. K. 1982. *An investigation into the formal and functional characteristics of qualifications in Legislative writing and its application to English for academic legal purposes.* a Ph.D. thesis, University of Aston in Birmingham, U.K.

Bhatia, V. K. 1993. *Analysing Genre — Language Use in Professional Settings.* London: Longman, Applied Linguistics and Language Study Series.

Bhatia, V. K. 1994. "Generic integrity in professional discourse". In *Text and Talk in Professional Contexts.* Britt-Louise Gunnarsson, Per Linell and Bengt Nordberg (eds), ASLA:s skriftsrie 6, Uppsala, Sweden.

Bhatia, V. K. 1995. "Genre-mixing and in professional communication: The case of 'private intentions' v. 'socially recognized purposes'". In Paul Bruthiaux, T. Boswood and B. Bertha (eds), *Explorations in English for Professional Communication.* Department of English, City University of Hong Kong, Hong Kong.

Bhatia, V. K. 1997. "Genre-Mixing in academic introductions". *English for Specific Purposes*, 16(3): 181-196.

Caldwell, Rchard 1982. Specialist informant interviews, reported in Bhatia 1982.

Fairclough, N. 1989. *Language and Power.* London: Longman.

Fairclough, N. 1993. *Discourse and Social Change.* London: Polity.

Featherstone, M. 1991. *Consumer Culture and Postmodernism.* London: Sage.

Ghai, Yash. 1997. *Hong Kong's New Constitutional Order: The Resumption of Chinese Sovereignty and the Basic Law.* Hong Kong: Hong Kong University Press.

Government of Hong Kong, 1984. *Sino-British Joint Declaration 1984.*

Government of Hong Kong, 1990. *The Basic Law of the Hong Kong Special Administrative Region of the People's Republic of China.*

Martin, J. R. 1993. "A contextual theory of language". In *The Powers of Literacy — A Genre Approach to Teaching Writing,* Pittsburgh: University of Pittsburgh Press (116-136).

Miller, C. R. 1984. "Genre as social action". *Quarterly Journal of Speech.* 70: 151-167.

Renton, Sir David 1975. *The Preparation of Legislation.* London: HMSO.

Swales, J. M. 1990. *Genre Analysis - English in Academic and Research Settings.* Cambridge: Cambridge University Press.

Conflicts in Professional Discourse: Language, Law and Real Estate

BIRGITTE NORLYK
University of Southern Denmark

This study focuses on two issues within the field of professional discourse. The first issue concerns the problem of interpretation of a sales contract drafted by one discourse community and modified by another. The second issue centres on the complex interaction between professional discourse, professional communication and the specific social stage on which professional discourse is enacted.

In connection with the selling of property in Denmark, the Seller's liability is normally regulated in a standard sales contract drafted by the Danish Board of Estate Agents. It is standard procedure for lawyers to copy this contract when they set up the actual deed. Due to long established work routines, the sales contract and the actual deed generally contain identical texts and both documents are considered legally binding (Andersen Krüger, P. et al. 1995).

Both the sales contract and the deed are characterized by a set of standard legal phrases dominated by the traditional complex syntax and the specialized terminology surrounding the legal community. While legal discourse may pose a problem in non-expert communication, it is generally considered to facilitate communication between members of the legal profession, since the specialized terminology and the complex syntax of legal discourse are part of the linguistic identity of the legal profession.

In 1988, the Danish Board of Estate Agents revised its standard sales contracts. The Board wished to update the somewhat complex language of the standard sales contract in order to facilitate communication with non-expert customers. As a result of a linguistic revision, part of a standard phrase was left out. According to the Board, this particular part of the phrase was redundant and added an unnecessary amount of legalese to the sales contract. Also, the Board felt that the archaic quality of this particular phrase stood in the way of clear customer communication. The revision of the standard contract thus introduced a rewrite of a particular standard phrase about the Seller's liability and the Seller's duty to inform the Buyer of possible, hidden flaws of the property in question.

Before 1988 the standard sales contract (see example 1) contained the phrase 'known to the Seller' in the section dealing with the Seller's liability and the Seller's duty to inform the Buyer about the state of the property. The Board felt that the words 'known to the Seller' were redundant since the Seller, logically, cannot inform a potential Buyer of flaws *unknown* to the Seller. The omission of the words 'known to the Seller' later caused great confusion at the Danish courts as concerns correct legal interpretation of the Seller's degree of liability. Subsequently, in 1994, the modernized version of the contract was replaced by the old version and its traditional legalese as illustrated in examples 1 and 2.

Example 1: Standard contract, old version.
(Used before 1988, and after 1993)
> ... *sælger erklærer, at de på ejendommen værende bygninger, sælger bekendt, er lovligt opført, indrettet og benyttet... etc.*
> ... the Seller declares that, known to the Seller, the buildings on the property are built and fitted in accordance with the proper laws and regulations...

Example 2: Standard contract, modified version.
(Used between 1988 and 1994)
> ... *sælger erklærer, at de på ejendommen værende bygninger er lovligt opført, indrettet og benyttet... etc.*
> ... the Seller declares that the buildings on the property are built and fitted in accordance with the proper laws and regulations...

Examples 1 and 2 later presented serious problems of interpretation in the Danish legal community as concerns the question of the Seller's liability. Do the two versions represent the same degree of liability or does example 2 represent a higher degree of liability and guarantee on the part of the Seller since the words 'known to the Seller' have been left out?

In the early 90s two cases were taken to court on this issue. In both cases the Buyer claimed increased liability on the part of the Seller as the words 'known to the Seller' did not appear neither in the sales contract nor in the deed. Both Buyers argued that the absence of the words 'known to the Seller' indicated a higher degree of liability and guarantee on the part of the Seller.

Although the cases are identical from a legal point of view, the Higher Courts were unable to agree on a correct legal interpretation as concerns the Seller's liability in the new version of the sales contract. Did the absence of the words 'known to the Seller' indicate a higher degree of liability on the part of the Seller? Or were the old version and the new version identical?

This question of interpretation still causes a certain amount of confusion in the Danish legal community. Consequently, the old version of the sales contract containing the words 'known to the Seller' was reintroduced in 1994. Meanwhile, the question of correct interpretation of the absent words remains unanswered in the legal community as illustrated in the following case stories.

Case stories

The first case story concerns the buying of a flat. A young student buys a flat from a lawyer in 1991. A few months later, the student is contacted by the person living in the flat below. She complains that water from the student's bathroom leaks into her bathroom.

The student duly contacts the Seller of the flat to complain about the problem. The Seller denies any knowledge of the problem. According to the Seller, the bathroom was modernized years ago long before he himself bought the flat and he consequently refuses to cover any repair costs. An expert later reports that the bathroom has been modernized by a do-it-yourself man and that some of the installations are illegal.

Referring to the section concerning the Seller's liability in the sales contract and in the deed, the student takes the Seller to court claiming that the modified version of the contract expresses an extended guarantee on the part of the Seller. The Seller, a professional lawyer, argues against this, claiming that the absent words of the modified version of the sales contract do not express an extended guarantee on his part.

The City Courts, however, decided that the new version did indeed express an extended guarantee on the part of the Seller, in spite of the fact that neither the Buyer nor the Seller had noticed the changed wording of the standard contract when the actual sale took place.

The second case presents similar problems of interpretation. In this case, the dispute between the Buyer and the Seller concerns electrical installations. In 1990, a few months after having bought an idyllic country cottage, the Buyer calls in an electrician who informs her that some of the electrical installations are illegal and need to be refitted.

The unhappy Buyer contacts the Seller who refuses to pay for the new installations. The Seller argues that he has never had any problems with the installations and that he did not know they were illegal. The installations were made years ago by an authorized electrician whose name, unfortunately, had slipped his mind.

Again, the Buyer takes the Seller to court. Referring to the new version of the standard sales contract, she argues that the changed wording of the modified version expresses an unlimited guarantee on the part of the Seller. Again, the City Courts agree with the Buyer: Compared to the old version of the standard sales contract, the new, modified version does indeed represent an extended guarantee on the Seller's part.

Conflicting interpretations

Both cases were subsequently appealed to the Higher Courts, whose judgments normally constitute a precedent for correct interpretation in the Danish legal community. In this case, however, the judgments of the Higher Courts failed to do so. Although the cases presented identical problems from a legal point of view, the Higher Courts failed to establish a framework for correct legal interpretation of the modified version of the standard sales contract for buying and selling property.

In the *first* case the three judges of the Higher Courts unanimously confirmed the judgment of the City Courts and interpreted the absent words 'known to the Seller' as indicating an increased degree of liability on the part of the Seller. Thus, in case one, the judgment of the Higher Courts was given *in favour of the Buyer*.

In the *second* case, however, the judges of Higher Court did not reach agreement on the correct legal interpretation of the absent words of the modified version of the standard contract. In case two, only one of the three judges of the Higher Courts agreed with the interpretation of the City Courts that the absent words of the modernized version did indeed indicate an increased liability on the Seller's part. Two judges of the Higher Courts, however, disagreed. In the professional opinion of these Higher Courts judges, the new, modified version of the sales contract did *not* express a higher degree of liability on the part of the Seller. According to these judges, versions one and two of the sales contract expressed the same degree of liability. Consequently, by majority vote, the Higher Courts in this case gave judgment *in favour of the Seller.* With one Higher Courts interpretation in favour of the Buyer and another Higher Courts interpretation in favour of the Seller, no precedent exists for correct interpretation of the modernized version of the standard contract.

In both cases the Higher Courts contacted the Board of Estate Agents to inquire about the Board's motive for changing the wording of the contract. The Board stated they had had absolutely no intention of changing the actual contents of the standard contracts. The changes in the wording of the contract had been made for *linguistic* reasons only in order to facilitate communication with non-expert customers. Apart from this comment the Board declined to discuss the matter any further and pointed to the fact that the old version of the standard sales contract had been reintroduced in 1994.

Basically, the question of correct interpretation remains unsolved in the Danish legal community. From a practical point of view, however, the problem has been solved as the standard sales contract is now back to its former version which includes the words 'known to the Seller'.

Discourse, communication and culture

Discourse communities with different cultures, different linguistic identities and different social contexts are brought together by everyday activities e.g. the activities surrounding the buying and selling of property. The cases stories illustrate that professional communities, like other types of organizations, have different attitudes towards communication (Miller 1995, Norlyk 1996). The interrelation between professional discourse and professional culture subtly influences the way professionals communicate both within their own community and outside of their professional community. Discourse, in other words, must be seen in relation to the social stage on which it is enacted (Bazerman 1993).

While the specific analysis of professional discourse is based on linguistic disciplines, the realization and interpretation of professional discourse take place in a complex context in which linguistic disciplines reflect only part of the total interaction. To analyse the context in which language operates we need an interdisciplinary perspective as argued a.o. by Schiffrin (1994) who claims that we cannot 'separate language from the rest of the world' and that we need to analyse further the relations between discourse and the social stage on which different types of discourse are enacted. 'To understand the language of discourse [.....] we need to understand the world in which language resides; and to understand the world in which language resides, we need to go outside linguistics' (Schiffrin: 418-19). When the Board of Estate Agents changed the wording of the standard contract to facilitate customer communication, it enacted a professional culture in which communication and customer orientation are highly valued. In contrast to the social stage of legal discourse, the discourse of estate agents is enacted in a social context in which the basis for continued existence depends on successful communication in a free market.

The importance of relating professional discourse to its cultural and social setting is stressed in a recent study analysing and comparing selected types of professional texts across 3 countries (Levin 1997). Like Schriffrin, Levin feels the need to establish a broader, multidisciplinary framework to explain the national differences in the structure of written professional discourse. In her comparative study of professional texts, Levin stresses that the structural, stylistic and thematic choices made by writers within a given profession are influenced not only by such factors as immediate context and genre but also by culture bound norms and socialization processes (Levin: 83-86).

Seen in this perspective, the case stories illustrate how the linguistic choices of a professional culture rely on the values and interests of the individual profession and the environment in which the profession operates. The socialization processes surrounding individual professional cultures subtly establish a framework for professional and linguistic identity. In a case story perspective, the legal community has one set of culture bound values, the community of estate agents another. To a large extent, the legal community and the community of estate agents operate in different worlds and in different contexts and share neither linguistic preferences nor professional identities. However, these different communities meet and interact in certain situations, e.g. in connection with the buying or selling of property.

In this situation the average citizen, generally a non-expert in the field of buying and selling property, finds himself caught between two types of professional communities with different types of discourse and different attitudes towards communicating with customers and non-experts. The discourse of the legal community and the discourse of estate agents reflect the different interests and the different cultures of each professional community. Both communities have distinct linguistic and syntactical preferences when it comes to communicating with customers, clients or non-experts.

While legal discourse has been the object of several studies concentrating on e.g. specialized terminology, syntax, and stylistic choices, little research - if any - has been done on the discourse of estate agents. Legal discourse constitutes a familiar area of LSP research. Legal discourse has been described a.o. as a type of LSP favouring archaic expressions, complex patterns of syntax, specialized terminology, seemingly endless repetitions etc. While it has been argued that this type of discourse or jargon may serve to monopolize knowledge or to establish the identity of a subculture (Morgan 1986, Charniawska-Joerges 1994), scholars within linguistics argue that the overall purpose of this type of discourse is to promote clarity and certainty within the professional community (Bhatia 1993).

The focus of communication in this complex type of professional discourse centers on the expert reader to whom this type of discourse expresses 'precision, clarity, unambiguity and all-inclusiveness' (Bhatia 1993: 102). The complex, professional discourse generally favoured by the legal profession conveys an impression of competence and professionalism to the expert reader. Depending on individual and situational factors, the non-expert reader may experience legal discourse as highly problematic. When the Board of Estate Agents changed the wording of the standard contract they improved non-expert communication. The linguistic revision performed by the Board reflects the social reality of a professional culture that depends on successful customer communication for its survival in the market. The revision of the contract, however, also demonstrate the Board's lack of awareness of the social reality surrounding the legal profession and its discourse. In contrast to the social reality surrounding estate agents, the legal profession does not depend on successful non-expert customer communication for its continued existence. Consequently, while the linguistic revision of the sales contract made perfect sense in the context of non-expert communication, the revision caused major disturbances in the legal context of expert communication.

Studies on legal discourse have demonstrated that specialized terminology and complex patterns of syntax constitute only part of the communication problem between experts and non-experts. In many cases pragmatic factors are equally important in successful non-expert communication. Thus, the non-expert reader's or receiver's degree of *prior knowledge* of the subject in question has been found to be of the utmost importance in professional communication with non-experts (Renkema 1993, Gunnarsson 1984).

In contrast to the fairly homogeneous legal community, in which members basically share the same educational background, estate agents constitute a less homogeneous group. Until recently, estate agents needed no formal qualifications although certain training courses did exist. The different educational backgrounds of estate agents may influence the concept of professional culture and professional identity negatively. However, all estate agents share one important condition. They all operate in a market context which demands a high degree of linguistic awareness and communication skills.

The discourse of estate agents

While legal discourse has been the object of many studies, the discourse of estate agents has received little attention within the field of discourse studies. Because of the nature of their work, estate agents need to be familiar with relevant aspects of at least two types of LSPs: that of the legal community and that of the banking or financial community. Also, seeing that they operate in a free market, estate agents need to pay attention to successful communication with customers and non-experts. Consequently, linguistic competency plays an important part in the social reality of estate agents. The changed wording of the standard sales contract is just one example of estate agents' awareness of the interaction between discourse and communication. To facilitate customer communication, estate agents a.o. have developed a specific type of discourse for describing property. To a certain degree this type of estate agents' discourse shares some characteristics of politically correct discourse in which expressions associating negative qualities must be avoided. In politically correct language, negative concepts like 'losses' or 'lies' undergo a terminological transformation and become 'negative cash flows' and 'terminological inexactitudes' (Beard and Cerf 1992: 134).

It may be argued that politically correct language and euphemisms characterized by the 'carefully chosen perspective' (Bülow-Møller 1994: 40) form an essential part of estate agents' discourse as regards communication with customers. First, estate agents need to present a given property in a way that appeals to the customer. Second, estate agents need a type of discourse which conveys basic information very quickly. Third, estate agents need a type of discourse which facilitates communication with non-specialist customers.

In a Danish context, the preferred media for representing this type of discourse is the local Sunday newspaper. Here readers find page after page advertising houses or flats in a highly standardized way that enables the individual reader to scan a page for property of potential interest to him. Graphic symbols such as pictograms symbolizing number of rooms, size of lot, distance to schools, and distance to shopping centers etc. are important in the visual discourse of estate agents, as graphic symbols convey information faster than verbal descriptions.

The fact that communication must be instant to catch the customers' interest is also reflected in the set of standard phrases favoured by estate agents in their descriptions of different types of property. While the legal community traditionally uses highly specialized legal terminology even when communicating outside of their professional framework, estate agents mainly use normal language when communicating with customers. However, the customer quickly discovers that she/he is actually presented with a special 'politically correct' professional code disguised as normal language. Pragmatic factors only enable the customer to decode correctly the special code of estate agents. After a few disillutioning visits to the properties advertized, the customer quickly learns to translate the discourse of estate agents into bleak reality.

Potentiel customers quickly learn to interpret correctly the inherent dichotomy of estate agents' discourse. Thus the phrase 'centrally located' often translates into located near motorway or railway junction while the phrase 'close to nature' translates into isolated or remote.

As concerns the actual state of the property advertised, the following phrases often appear: Some houses are described as having 'lots of potential'. Others 'need love and care' while others again represent the perfect buy 'for the do-it-yourself man'. In all three instances pragmatic factors instantly enable the reader to decode the above descriptions as meaning that a fair amount of money and hard work need to be invested to make the house or the flat habitable.

Apart from the legalese of sales contracts, the discourse of estate agents contains only few examples of what we traditionally understand by specialized terminology and professional discourse. On the other hand, the discourse of estate agents represents a special type of professional discourse that needs to be decoded or translated by the reader. The key to this decoding consists of a mixture of pragmatic factors and a good deal of creative cynicism. From a communication point of view non-experts quickly learn to decode this special type of professional discourse, while the decoding of legal discourse remains a problem in non-expert communication since pragmatic knowledge is harder to come by.

As demonstrated in the case stories, the linguistic identity of individual professional cultures will naturally differ. The social stage of a profession determines its norms, its values, and its discourse. In everyday life, the different discourses of lawyers and estate agents meet in connection with the buying and selling of property. The case stories illustrate how differences between professional cultures and their linguistic preferences may lead to unforeseen conflicts when established patterns of professional discourse are altered by non-professionals.

Conclusion

The legal community and the community of estate agents represent professional cultures that operate in different contexts and in different markets. The specific market situation of a given professional culture necessarily affects the attention paid to the discourse used in communication with non-expert customers or clients. Survival for the estate agent is directly connected to the way in which he communicates with his customers. If a text does not communicate in a non-expert context the text must be redrafted or changed as demonstrated in the Board's modified version of the standard sales contract.

The legal community, on the other hand, operates in a different social context in which the issue of clarity and precision takes precedence over external communication. In the wider social context surrounding the legal profesion, the complex syntax and specialized terminology of legal discourse may also serve a different set of purposes. Apart from establishing a sense of professional identity and clarity, the arcane quality of legal discourse discreetly serves to uphold the power and influence of the legal community (Conley and O'Barr 1998: 6-14, 134-135).

In everyday life the meeting of different professional discourse communities with different professional cultures may result in serious communication clashes as demonstrated in the case stories. Research on organizations has shown that specialized vocabulary, professional jargon and culture bound interpretation systems constitute an important part of the organization's identity and affect the total efficiency of the organization (Smircich 1992). This paper argues that the same mechanism applies for professional communities. In this particular case, the traditional interpretation pattern of the legal community was deeply affected when the Board of Estate Agents introduced changes in long established, culture bound legal discourse.

Originally, the *motive* for the changed wording of the sales contract was to facilitate non-expert customer communication. However, the Board failed to recognize the potential danger of changing another profession's specialized discourse and involuntarily created a communication paradox. While non-expert communication was improved, expert communication within the legal community was affected negatively. As the legal community was unable to present a unanimous interpretation of the modified version of the standard contract, the old version of the contract was reintroduced in 1994.

References

Andersen, Paul Krüger et al. 1995. *Dansk Privatret*. København: DJØF.
Bazerman, C. 1993. In N. R. Blyler and C. Thralls (eds), *Professional Communication: The Social Perspective*. Sage.
Beard, Henry and Cerf, Christopher. 1992. *The Official Politically Correct Dictionary*. Grafton.
Bhatia, Vijay K. 1993. *Analysing Genre: Language Users in Professional Settings*. London: Longman.
Bülow-Møller, A. M. 1994. *The View from the Bridge*. København: Samfundslitteratur.
Charniawska-Joerges, Barbara. 1994."Narratives of Individual and Organizational Identities". In S. Deetz (ed.). Communication Yearbook/17. Sage.
Conley, John M. and William M. O'Barr. 1998. Just Words: Law, Language and Power. University of Chicago Press.
Gunnarsson, Britt-Louise. 1984. "Functional Comprehensibility of Legislative Texts". *Text* no.4.

Levin, Anna. 1997. *Kognitiva och pragmatiska mönster i professionella texter från svenska, engelska och tyska skrivmiljöer.* TEFA NR 20. Uppsala Universitet.
Miller, Katherine. 1995. *Organizational Communication: Approaches and Processes.* Wadsworth.
Morgan, Gareth. 1986. *Images of Organization.* Sage.
Norlyk, Birgitte. 1996. "Miscommunication and discourse practices in occupational cultures". *International Journal of Applied Linguistics* 6(1).
Renkema, Jan. 1993. Discourse Studies. Amsterdam: John Benjamin.
Smircich, Linda. 1992. "Organizations as Shared Meanings". In J. Shafritz, and J. Steven Ott (eds), *Classics of Organization Theory.* Wadsworth.

Genres and New Technology
(Changing and emerging genres)

Powerpoints: Technology, Lectures, and Changing Genres

GREG MYERS
Department of Linguistics and Modern English Language
Lancaster University, Lancaster LA1 4YT
United Kingdom

Introduction

When I moved from the University of Texas to the University of Lancaster a dozen years ago, one of the first things I had to re-learn, along with driving, ordering in restaurants, and saying "sorry" instead of "excuse me", was giving lectures. In the US, I had always just prepared a list of notes and passages on, say, Faulkner's The Bear, and counted on students' questions and their responses to my questions to get me though the hour. And since much of the assessment was based on coursework, this kind of discussion, and the sense of engagement with the teacher, might serve the purpose of starting students off on a reasonable essay. In the UK, where most assessment is still by an exam based on the lectures, it is crucial, or they think it is, that they produce usable notes. The students neither asked nor answered questions in public. To do so would be to come across as something of a swot. Also, the lectures are bigger, with maybe 200 students rather than the 35 or so typical for a US class. Any discussion is left for seminars. And they wanted a clear, well-organized, practiced performance that fits the syllabus and the fifty minutes exactly.

At first I learned this form by watching the other lecturers. I was in the English Literature department then, and admired the old style lecturers who could sit at the front, rocking slowly forwards and backwards, reciting poems that I could never remember even on exams, and systematically moving through the headings on the handout. I could never pull off this donnishly casual approach. Even though I've never simply read out my lectures, in the US or the UK, I found that I had to start writing them out, including the jokes, so that I planned them carefully enough. So, though I've never studied lectures as a genre, my writing and delivering two or three a week has led me to think about their form. (For references to those who have studied them as a genre, see Thompson 1994). As I became more experienced, I began experimenting with some other ways of getting the information across: slides, videos, little tasks for the students to do.

In this chapter I would like to consider one of these experiments: I have started using Microsoft Powerpoint™ presentation software instead of overhead projector slides in one course. This may seem a small enough change — I thought so myself at first — but I found it raised wider questions about technology and genre that could be of interest even if you aren't concerned with lectures. I will use this example to raise more general questions about where genre analysis is and where it is going. Genre analysis includes now a wide range of researchers and teachers and a wide range of kinds of texts (for reviews, see Swales 1990, Bhatia 1993, Bazerman 1994, Berkenkotter and Huckin 1994, Connor 1996, Martin 1989). These very different researchers in linguistics, ESP, education, and rhetoric share three assumptions:

- Genres link the textual and the social
- Genres are relatively stable, because they are based on stereotypical social acts
- Genres can be analyzed in terms of participants' purposes

So the analyst looks for regular moves in texts, and correlates them with regular social practices and institutional structures. One application of this analysis was in examining academic research articles (Bazerman 1988, Swales 1990, Berkenkotter and Huckin 1994). By looking for the stereotypical social actions underlying these texts, researchers were able to show there was some sense underlying the complex conventions: the writers were "Clearing a Research Space" (Swales 1990) in a crowded and competitive field. Their aim was to use this knowledge to help teachers prepare students for writing these genres. It also gave writers a way of understanding variations between disciplines and between national academic cultures.

Because it focuses on the stereotypical and conventional, this framework is less useful in helping us understand how genres change. It tells us to look for purposes and interactions, but does not tell us why these might vary over time. And this problem does not just arise with a pedagogical genre like lectures; we can see changes in US job application letters (as they move towards more explicit self-assertion), or British television commercials (as they moved in the 1960s from monologues to dialogues), or tabloid newspapers in the 1920s, as they began to incorporate news photos. In each case we need to know, not only how to account for the change, but even how to describe it. At the end of this chapter I will raise some issues about how the case of lectures might relate to changes in other genres.

Changing genres: The old lecture

In the traditional lecture, I write down more or less what I want to say, and maybe prepare a handout that contains, say, some quotations, and an ad I want them to look at. I don't read this out, but keep the written text in front of me as a prompt, making asides as I go along. Students arrive at 9:00 on a Wednesday morning at George Fox Lecture Theatre 1, sit there facing me, and write something — who knows what — in their notes. A couple of months later, they sit in an examination hall, and they may use some of this information in their written answers. It is a kind of distorted quadrilateral, from my writing to my reading and speaking to the students' listening and the students' writing.

My Lancaster colleagues may be offended by this description, and the implication that everyone does such a traditional performance. They have pointed out to me that there are many other possible variations:

- in numbers of students — from 30 to 300, with appropriate changes in style
- in lengths — shorter or longer than the standard fifty minutes
- in formats for delivery — including discussion, exercises, and feedback
- in aims — not just presenting information, but engaging students, displaying enthusiasm, demonstrating examples.

They accused me of using the idea of genre to impose uniformity on these diverse practices: a common criticism of genre analysts.

Yet despite this diversity, we all think we know what a university lecture is. When I tell a colleague I have a lecture at 9:00, they think they know just what is involved without any further details. And the students have certain expectations, which one defies at one's peril. If you stop at 35 minutes past the hour instead of 50, you will be met with puzzled dissatisfaction, not delight. If you walk around the room, as you can do with a radio mike, you will be regarded with the kind of stares from students that I imagine urban gang members give the police.

And if, on the other hand, I gave one of my best course lectures at an academic conference, the audience of colleagues would know something was wrong. It might take a while to say just what was wrong, but my relation to them, my use of imperatives and deictics, my opening and closing, would be slightly off. (I have often wondered why speakers at academic conferences and visitors giving academic seminars never allowed their audiences a short break in the middle of a fifty-minute talk, as I do with course lectures. Perhaps the problem is that such a break would rudely imply that you have to pay close attention, and therefore are easily exhausted).

Genre analysts tell us to look for the source of this conventionality in some stereotypical social action (Miller 1984). Some of the odd features of course lectures can be explained in terms of the purposes of teachers and students, and the transformation from writing to speech to writing again. I want to engage them enough so that they take an interest in the topic; they want to write down the right things for the exam. Course lectures are streams of words that are intended to turn into the blocks of notes on the student's page. So the first requirement of a good lecture, we learn in our training courses, is clear signaling and constant review, keeping the string of the message connected and untangled. There can be asides, breaks, examples, but these shifts of level or direction must be clearly signaled, because the experience of the lecture is one of sequence. Since it takes place over a set time, the lecturer's text must be timed to fit. It needs some sort of engaging opening — such as a striking example — some sort of break in the middle, since their attention doesn't last an hour, and a review at the end. It must be enormously redundant, because even the most attentive student, with the most fascinating lecturer, is likely to find that their mind wanders for part of the lecture. And this sequence is part of the larger sequence of the course and the degree, so the lecturer must constantly make these links as well. This is what is usually taught in staff development courses, on the lecturer's side, and study skills courses, on the students'.

What is so essential about speaking and hearing the words? In the middle ages, it was the cheapest way of getting one text to a number of students. Now I could just photocopy it (which I don't do) or save it in HTML format and put it on the World Wide Web (which I do). So why should students turn up? There have been several arguments given by defenders of lectures over the generations.

- Students learn better by listening, selecting, organizing, writing down, and reviewing. (There may also be a discipline for the lecturers in writing them, though I haven't heard anyone urge that as a reason in their favor).
- Lectures allow for demonstrations, whether experiments in chemistry, readings in poetry, or my video advertisements.
- More subtly, the co-presence of student and lecturer allows for much more complex kinds of information, for instance verbal and non-verbal emphasis, or de-emphasis, or marking of asides from the main text. (This was brought home to me when I had a biologist lecture in my course, and was astonished

by how much she marked information). More subtly, the lecturer is signalling attitudes and engagement.

- Co-presence also links to the disciplinary function of lectures. I can see the students, and they can see that I can see them, and they know I can see if they are not there.

So lectures may make some sense if we think about them as interaction. Still, it is, when one thinks about it, a rather strange kind of communication. Three kinds of changes have brought this home to me; they are the sorts of changes that make a practitioner of a genre begin to step back and analyze it. One I've already mentioned, the kind of relearning I had to do when I came to Britain. Another change that made me think about the lecture is that we have been preparing for the possibility of deaf students on the course. One should, of course, have sign language interpreters for them. Or one can have designated note-takers, who take the notes for the deaf student. At a workshop, an advisor for deaf students pointed out that sometimes they would just prefer to have the lecture notes. This is what finally brought home to me the oddity of my having a written text, which I read out, and they write down, producing a garbled form of my notes. And if it makes sense just to hand it to deaf students, why not just hand it to hearing students as well, and save them the trouble of getting up for a 9 o'clock lecture? A third change that makes the lecture look strange is that, since the lecture is about advertising, I use several modes besides my spoken words: usually several videos, overhead projector slides and projections from a visualizer, maybe tapes of radio ads, and sometimes projection of a computer screen. All these bits of performance make it preferable for students to attend the lecture, rather than getting notes from a fellow student. But they also mean that the speaking and writing are just one of the things going on. Why shouldn't it all be on tape, like an Open University programme?

All this makes it sound as if all the reasons for change in lectures were internal, based on the logical development of a form of communication. But one could also trace the pressures for change in terms of change in terms of the transformation of institutions, the shifting boundaries of social worlds, or the impact of new technologies:

- growing student numbers and not enough space
- increasing staff teaching hours
- the distance of students from the university
- admission of students not prepared for the lecture system
- students' feedback on the course, and course choices
- changing ideas of the knowledge that is to be gained

- changing forms of assessment
- availability of new projection, networking, and multi-media equipment.

Thus there is a very powerful tradition stabilizing the genre, but also pressures for change.

The new lecture

I am going to look at just one change in how I give lectures, specifically, in how I present material on the overhead projector. As well as struggling with current social theory and developments in discourse analysis, I have been wrestling with Microsoft Powerpoint™. This is a piece of software, included in the Microsoft Office package, meant to help business people with the visuals for their presentations, especially presentations that may be repeated often by different people, such as those in sales, training, or planning. It is like a word processor, but it produces a series of slides for projection, either with an OHP or directly onto a monitor or screen. It also produces an outline, and handouts that can contain notes. I have always wanted to try it out, since I saw it used by a very good lecturer on our course. In any case, I was already putting more and more of the detail of my lectures onto overhead projector slides, partly because students complained that if too much was on the handouts, there was no point in getting up for the lectures. (That's an important point — some of the major shifts in technology are acceptable because they are seen at first as extensions of existing technology).

It might seem that this was a very minor change in my writing of lectures. Most of what Powerpoint™ does can be done, albeit more laboriously, with any word processor. As with many technology innovations, it can apparently be reduced to its predecessors. I could use Word to put various pages in big type and print them out; I could even scroll through them on a screen. Powerpoint™ just links a number of functions at once, so as one prepares an outline one is also making the basis for slides and a handout and notes. It also allows one to change one's slides up to the last minute, instead of printing them out and having them set in plastic. As with other Microsoft products, one main claim is that the work in one product can be transferred to another, so Powerpoint™ work can be transferred to Word or saved in Internet Assistant for a Web Page.

But Powerpoint™ is not just another way of doing what one is already doing more efficiently; one cannot use it without realizing that it will shape one's presentation, subtly or not so subtly.

- The most important effect is that it gets you to think of your talk in terms of slides; it becomes a series of blocks rather than a string.
- Within each block, it gets one to think of a hierarchical series of sections and bullet points.
- It calls for certain kinds of visual material, a certain format, and it encourages some visual material wherever possible
- It puts a corporate identifier on each slide, and encourages conformity to a visual style.

I first realized how much Powerpoint™ defined one's form when I saw they had, already set up, a format for presenting "Bad News". When you choose "Use a Wizard", it gives you a talk already formatted for genres like this or Sales, or Report. Of course business letter and public speaking guides have given such formats since the middle ages — but here they are built into the writing tool. One can instead choose templates (the type and background) and layouts (the arrangement on the page) separately, and build up one's talk from scratch. But even then one is constrained by the kinds of layouts. For instance, one can have centred title text, and it will prompt you for a sub-title. Or you can have a title above a bullet-point list, or a title above two columns. But you can't have a title over three columns of text, or a picture by itself. Within each page, one is limited to a certain size print, a certain number of points, and almost infinite but carefully related systems of colors of text and background. All this is quite sensible: there are good reasons to use yellow text on a blue background, or have no more than four points on a slide, or always have a review at the end. My point is that these general rules of thumb are already encoded in the tool.

All this, you might think, is just the surface, not the deeper form or content of a presentation. But just as Charles Bazerman (1988) showed the APA Style Sheet defined a certain rhetoric, Powerpoint™ defines a rhetoric, one based on a hierarchical outline, leading to an appearance of logic. It also draws on the visual as attention getting, soothing "eye candy" rather than as illustration of the text. As I took my old lecture on Advertising Regulation and rewrote it for Powerpoint™ presentation, I made the following changes:

- I decided it had a sub-title
- I put all the examples of disclaimers on one slide as the introduction, instead of a story about a current case
- I broke the section on history into separate periods, creating parallelism
- I picked out some bullet points - short points to fit on a line, such as names and dates

- My three categories of the talk (Laws, Agreements, and Customs) melted down into two when I found the third didn't neatly fill a slide
- Laws had to be made into three slides, one with the uninspiring title of "More Examples of Laws"
- I realized I had enough comments in separate sections for a new slide on "European Regulations".
- Other new slides were on "Sex" and "Political issues"; in each case the creation of bullet points made me rethink the headings.
- Where I had too many bullet points, I sometimes made the last one into a new slide and discussed it at greater length
- I eliminated long sections at the end that didn't seem to break down into bit sized chunks (anyway, I could see I would run out of time)
- I tried to include some of the ads in the slide show, but I didn't scan them in the right graphics format.
- I ended, as all wizards do, with a review of main points, rather than on the provocative case I had used before. I realized as I summarized them that they were rather different from what I thought they were.

The overall effect is that what was before a carefully connected sequence, with some digressions for stories, and references to texts on a handout, was now a series of spaces, marked by rather flashy transitions.

But these lists of formal changes don't quite get at the shift in effect. One clue to the shift came from a student who told me that with so much attention to the screen, he didn't pay attention to anything I said that wasn't up there. Another clue came when I saw a rough cut of a promotional film that included some shots of my lectures. I am backlit, my face in silhouette against the bright, colourful words on the screen. It looks good as a video effect, but it also shows that the written text, produced by the machine, has become the star; I am reduced to an unseen voiceover of my own lectures. That may not matter in a business setting, where different people from sales or personnel may be called upon to speak the same words. But for a university lecturer, it marks a shift in what Goffman (1981) called *footing*; that is, I am seen as the animator rather than the source of the utterance. Instead of my speaking with the aid of some visual device, the text is speaking with my aid.

Let's review then some of the shifts I have mentioned. First, and most obviously, there is a shift in the text form, from more sequential to more hierarchical. There is a shift in the process of presentation, from an emphasis on continuity to an emphasis on chunks. There is a shift in the kind of interaction, from voice to text. Finally, and most subtly, there is a shift in the framing of knowledge. If lectures are a form of authority, a long monologue, Powerpoint™ presentations are more obviously a form of rhetoric. The software was designed for persuasion, not exposition or debate.

Changes in genres

The case of Powerpoint™ illustrates five big issues that seem to arise in much of current work on genre, and seem to be likely avenues for further theoretical and empirical exploration.

- Technology and the material form of texts
- Time and the experience of texts
- Multiple authors and audiences
- Institutions and boundary work
- Embodied knowledge and skill.

Under each heading, my experience with this bit of new technology highlights issues that are being discussed in all sorts of other cases across social theory.

Technology and the material form of texts

I am not at all saying change is a matter of technological determinism, that Powerpoint™ made me write this way. After all, I can always turn it off, or more likely, never plug it in. Rather, the technology by its very new-ness reminds us of the material nature of all texts. We are constrained in any lecture by A4 pages, lecture theatres, OHP slides, ASCII characters, black and white photocopies, video formats, even blackboards and whiteboards.

Analysts all know the importance of material form because of their own involvement with writing and publishing, but we have been slow to take on multi-modal texts (see however, Kress and van Leeuwen 1996, and Faigley 1997). We don't yet have the analytical tools for these that we have for words. Our students may be more familiar with these other modes — in Web pages and desk top publishing — than we are.

Time and the experience of texts

On one level, Powerpoint™ is an ideal case for reminding us of the role of time in genre. It unveils its slide show in time, in fancy transitions between slides and builds between lines. In a larger sense, it offers the organization more control over time: it can provide on a disk a flexible and permanent framework for presentations, it can structure the way that the representative and the audiences interact, and it can even run automatically without the presenter present, at trade fairs or open days.

But it also shows up the difficulties that genre analysis has in dealing with time. Genre analysis is based on *stereotypical* action, repeated and paralleled. It involves the removal of time from texts, so it increases the difficulty of studying texts in terms of the experience of time and change. Ethnographic approaches that take us into the enactment of texts (such as Prior 1994 or Winsor 1996) can incorporate the time of delivery, of repetition, of revision.

Multiple authors and audiences

Genre analysis based on ideas of purpose has had some trouble dealing with multiple goals, as in patent applications for example. Even within one actor there can be various purposes in tension. Lectures have always had heterogeneous voices; mediated through my voice as lecturer, students hear a mixture of voices from critics, ads, the Advertising Standards Authority, historians. In the Powerpoint™ version of the lecture the author is often seen as the organization whose logo appears on each slide (the set-up Wizard at the beginning prompts the writer for an organization name). The unity of the source is stressed visually with this logo or company name, even where it is not clear who is speaking. The performance is standardized, but it is also removed from its immediate context, made repeatable for different audiences by different speakers. Of course, as several people have pointed out to me, this standardization can be resisted, simply by using the software in different ways, opening up screens to write out student feedback, or marking on the slides. The technology makes it easier to do some things than others.

Institutions and boundary work

Genre analysis started with neatly defined special purposes such as medical interactions or the research article. What we have seen recently is the blurring of these boundaries: between academic and commercial, specialist and popular, student and consumer, national and international (Engeström, Engeström and Kärkkäinen 1995, Myers 1996).

My use of Powerpoint™ crosses another boundary. After all, why do we have such fancy software, taking up many Mb on our disks? I don't imagine any academics asked for it; most of us started with chalk and have just figured out overhead projectors. It was designed for business presentations. I used to joke that you could tell the non-academics at any conference — they had the coloured slides. What we are seeing is a style of presentation, a technology of discourse (Fairclough 1992), carrying over from the world of business to that of academics. It parallels the way word-processing entered academic routines from business, or the web went from academic applications to business. This is no conspiracy of corporations to dominate academic lectures (about which they care little) but an unintended effect of the way software becomes available.

Embodied knowledge and skill

Swales and Bazerman explored genre analysis as a way of tracing knowledge-making in research articles. More recent work has looked at other kinds of stabilization, in patents, inquiries, manuals, handbooks, reports (see Myers 1994, Bazerman 1997, Trosborg 1997, Winsor 1996). In the case of Powerpoint™ presentations, we see that information is presented rhetorically, strategically, oriented to the use the audience can make of it and the action they are expected to take. I thought at first that such lectures were stripping out the context of interaction, but what is going on is that it is embedded in another context, that of business.

Conclusion

Throughout this chapter I have been stressing several features of the change in genre made with Powerpoint lectures: that the changes were apparently small ones, merely formal; that the deeper shifts were unnoticed until I became aware of students' responses; that the changes were related to broader changes in the lecture form and in the provision of university teaching; and that they involved carrying an artifact — the software, from business to education. All these issues lead us back to the pun in my title; where is the power behind this change?

But it is easier from this case to say what the power is not. The trademark Powerpoint™ suggests that you the user will be empowered and more focused, but this is not the case because the software can be used (but need not be) to de-skill the presenter. The power is not in Microsoft, as personified in Bill Gates; they did not try to get me to take up this product, and indeed clearly developed it for other uses. The teachers seem to have power over the students, in preparing the lectures and the exam, but they find themselves driven by students evaluations or by their more subtle indifference. Things seem to have power — the computer, the lecture room, the notebooks. But they do not act apart from human strategies.

A more useful strategy is to stop looking for the power, and to see complex purposes and strategies in each actor (Myers 1996). Action involves enlisting others, so I enlist Microsoft business software, Microsoft enlists educators, the software enlists my voiceover. But none of these actors is locked in deterministically; these links can come apart. We both use genres and are channelled by them; that is the tension that must guide any of our analyses, whether of corporate annual reports, handbooks, e-mail, newspapers, or leaflets. Change comes, not from the inside or outside, but in that tension.[1]

Note

[1] My thanks to all those who commented on this presentation, at the conference at Aarhus, especially Paul Wickens and Charles Bazerman, and also those at presentations of other versions at the Linguistics Department at Lancaster University and the English Department at Indiana University Purdue University Indianapolis. Thanks also for extensive written comments from Carol Berkenkotter and Lester Faigley. My thinking about genre has been influenced, not only by those works cited, but by a number of recent PhD dissertations that are not yet in print. So I would like to record a debt to Kevin Nwogu (Aston), Nshindi Mulamba (Lancaster), Deo Ndoloi (Lancaster), Mary Muchiri (Lancaster), Sanne Knudsen (Roskilde), John Holmes (Lancaster), Susan Thompson (Liverpool), Sally Burgess (Reading), Solbjorg Skulstad (Bergen), Simon Pardoe (Lancaster), Christine Räisänen (Gothenberg), Colin Barron (Lancaster), and Chris Tribble (Lancaster) — with apologies that I can no longer trace just what bit of my development I owe to whom.

References

Bazerman, C. 1988. *Shaping Written Knowledge: The Genre and Activity of the Experimental Article in Science*. Madison: University of Wisconsin Press.
Bazerman, C. 1994. *Constructing Experience*. Carbondale: Southern Illinois University Press.
Bazerman, C. 1997. "Performatices constituting value: the case of patents". In B.-L. Gunnarsson, P. Linell, and B. Nordberg (eds), *The Construction of Professional Discourse*. London: Longman, 42-53.
Berkenkotter, C. and Huckin, T. 1994. *Genre Knowledge in Disciplinary Communication*. Hillsdale, NJ: Lawrence Erlbaum Associates.
Bhatia, V. 1993. *Analysing Genre: Language Use in Professional Settings*. London: Longman.
Connor, U. 1996. *Contrastive Rhetoric*. Cambridge: Cambridge University Press.
Engeström, Y., Engeström, R., and Kärkkäinen, M. 1995. "Polycontextuality and boundary crossing in expert cognition: learning and problem solving in complex work activities". *Learning and Instruction* 5: 319-336.
Faigley, L. 1997. "After the essay: the literacy of color, sound, and motion". http://www.dla.utexas.edu/depts/drc/faigley/after.essay/after.essay.html.
Fairclough, N 1992. *Discourse and Social Change*. Cambridge: Polity.

Goffman, E. 1981. *Forms of Talk*. Philadelphia: University of Pennsylvania Press.

Kress, G. and van Leeuwen, T. 1996. *Reading Images: the grammar of visual design*. London: Routledge.

Martin, J. 1989. *Factual Writing: Exploring and Challenging Social Reality*. Oxford: Oxford University Press.

Miller, C. 1984. "Genre as social action". *Quarterly Journal of Speech* 70: 151-67.

Myers, G. 1994. "From discovery to invention: two researchers write patents". *Social Studies of Science* 25: 57-105.

Myers, G. 1996. "Out of the laboratory and down to the bay: writing in Science and Technology Studies". *Written Communication* 13: 4-43.

Prior, P. 1994. "Response, revision, disciplinarity — a microhistory of a dissertation prospectus in sociology". *Written Communication* 11: 483-533.

Swales, J. 1990. *Genre Analysis*. Cambridge: Cambridge University Press.

Thompson, S. 1994. "Aspects of cohesion in monologue". *Applied Linguistics* 15: 58-75.

Trosborg, A. 1997. "Contracts as social action". In B.-L. Gunnarsson, P. Linell, and B. Nordberg (eds), *The Construction of Professional Discourse*. London: Longman, 54-75.

Winsor, D. 1996. *Writing Like an Engineer: A Rhetorical Education*. Hillsdale, NJ: Lawrence Erlbaum Associates.

Powerpoints:

Technologies, Lectures, and Changing Genres

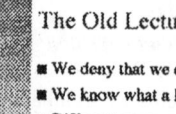

The Old Lecture

- We deny that we do typical lectures
- We know what a lecture is
- Different purposes
- Interaction in lectures
- A defense of lectures

Why lectures are changing

- Experience of visual and text media
- Institutional time pressures
- Audience as students or consumers
- Basis in discipline or topic
- Content as skills, knowledge, experience

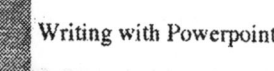

Writing with Powerpoint:

- Think in terms of slides, not strings
- Think in a hierarchy of sections, not sequence and asides
- Integration of certain kinds of formatted visual material, not pointing to handout
- Includes corporate identification

Standardization in the software

- Wizards (Sales, Bad News, Training . . .)
- Templates (font, colour, and background)
- Layout (text and graphics)
- Outline (hierarchies)
- Convertability

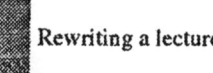

Rewriting a lecture

- added subtitle because there was a slot
- put examples on one slide
- divided history into separate slides
- pulled out names and dates as bullet points
- Rephrased points for parallelism

More on rewriting

- Made new slides where there were too many bullet points
- melted three categories down to two
- Eliminated long sections
- Tried to include some ads
- Ended with Main Points

Shifts in the genre

- Shift in form from seqential to hierarchical
- Shift in performance from continuity to chunks
- Shift in interaction from voice to text
- Shift in knowledge from authority to rhetoric

Analyzing changes in genres

- Technology and the material form of texts
- Time and the experience of texts
- Multiple authors and readers
- Institutions and boundary work
- Embodied knowledge and skill

Main Points

- Genres are based on social action
- The changes with Powerpoint are apparently small ones
- The changes are related to broader changes in university teaching
- they involve carrying an artifact from business to education

Rhetorical Clustering and Perceptual Cohesion in Technical (Online) Documentation

LARS JOHNSEN
Southern Denmark Business School

Introduction

One noticeable characteristic of many, if not most, documents used for technical communication purposes these days is that they constitute intricate webs of digital information in which resources such as text, layout, typography, tables, pictures and even animation sequences and video clips all play an increasingly significant role.

Although, of course, I do not think that LSP genre analysis can, or should, take all these aspects into consideration, I do believe it is worthwhile sometimes, not least in educational contexts, to examine more closely how professional communicators such as technical writers achieve communicative goals in specific genres not only through language but also through the interaction of text and visual design.

One aim of genre-oriented text design analysis might be to investigate how, or to what extent, communicative intentions, or semantic structure elements (Halliday and Hasan 1989), are realized by what Schriver has called *rhetorical clusters*, Gestalts of text elements, verbal or visual, designed to work together as functional units of purposeful and related content (Schriver 1997: 343).

For instance, a genre analyst of software manuals might be interested to find out what design strategies technical writers employ to convey important information such as warnings or cautions and how they ensure that this information actually gets across to the user of the software. The analyst would find, as is shown in figure 1, that one trick of the technical writing trade is to create a rhetorical cluster by combining a foregrounded, often prefixed, lexical marker (a word like *note, caution* or *important*) with a layout which makes the warning or note stand out from the rest of the text. What happens here, in terms of perceptual effect, is that by partly boxing the information of importance the writer manages to create what is known as closure in Gestalt psychology, a stable, self-sustaining structure which is readily conspicuous to the reader (Moore and Fitz 1993).

Figure 1

Sharing files with earlier releases

A Frame product automatically opens documents created with an earlier release of FrameMaker (2.0 or higher).

To use an earlier release of FrameMaker (such as 2.1) to edit a document created with a later release of FrameMaker (such as 5):

1. Use the newer Frame product release to save the document in MIF.
2. Open the MIF file with the earlier release of FrameMaker.

Important: Earlier releases of FrameMaker do not support all MIF statements in the current release. For example, when you use release 2.1 of FrameMaker to open a document created in release 3 of FrameMaker, MIF statements describing tables and conditional text are skipped. Ignore the related error messages. For a description of the differences between MIF 5.00 and previous releases, see Appendix C, "MIF Compatibility."

Modifying documents
You can use MIF to perform custom document processing. For example, you can create a program or write a series of text editor macros to search for and change paragraph tags in a MIF file. You can also edit a MIF book file to easily add or change document names in a book.

For an example of using MIF to easily update the values in a table, see "Updating several values in a table" on page 215.

Example taken from FrameMaker® Online Manual

Aims

Below I shall try to focus on the relationship between text design and textual organization in technical documentation, especially online documentation for end users. In particular, I wish to:

a) exemplify how technical writers may organize rhetorical clusters in technical documentation through the interaction of text, layout and typography

b) discuss how we may describe the structure and texture of rhetorical clusters in digital documents on the basis of Campbell's principles of perceptual cohesion (Campbell 1995).

One reason why effective rhetorical clustering is so important in technical user documentation is that this type of text must often support a variety of functions: skimming, scanning, the retrieval of specific items of information, reading to learn, reading to do, etc. And clear textual clues are absolutely crucial in *online* documentation because of the inherent limitations of the computer screen as a communication interface. A computer monitor provides a smaller reading area than, say, a book; the resolution of text is poorer; and the reader is only presented with a "tunnel view", so to speak, of the contents of the document.

Text, typography and layout

An example of rhetorical clustering is seen in figure 2. This is an entry in an online reference manual which describes the copy function in a computer programme and how this feature is used. What makes this text hang together as a document is not merely its semantic cohesion mechanisms but also the way the technical writer lets design elements like parallelism, graphical symbols, typefaces and lines interact.

Thus, it is essentially the consistent use of white space, headings and horizontal lines which supports our segmentation of the text into five more or less self-contained clusters each with its own communicative function: information on where this menu command is to be found, what it may be used for, how it may be used, what alternatives are available and finally where related material may be found.

Figure 2

Copy
(Edit, Copy)

Overview

Copy allows you to quickly copy large sections of text or graphics within an infobase, to or from other infobases, and to or from other Windows applications. It does this by copying the text and/or graphics selected to be copied to the Windows Clipboard for temporary storage. The selection remains in the clipboard until another selection is cut or copied, at which time it is replaced.

Description

To copy a section from the current infobase:

Step 1 Select the text and/or graphics to be copied.

Step 2 Choose **Edit** and **Copy** from the main menu.
- You can also use CTRL+C from the keyboard to copy.
- The selected region is copied into the clipboard.

Step 3 Move the insertion point to the position you want the text copied to (within the current infobase, another infobase, or another Windows application).

Step 4 Select **Paste** from the **Edit** menu.
- You can also use CTRL+V from the keyboard to paste.
- The selected area is copied at the new position.

Quick Keys

Copy: CTRL+INSERT or CTRL+C
Cut: SHIFT+DELETE or CTRL+X
Paste: SHIFT+INSERT or CTRL+V
Undo: ALT+BACKSPACE or CTRL+Z

Cross References

For additional or supplementary information, see:

> *Clipboard*
> *Cut*
> *Paste*
> *Paste special*

Example taken from Folio Views® Online Reference Manual

In the third cluster - the unit entitled "Description" - it is primarily the interplay of linguistic and graphical parallelism which helps us perceive continuities and discontinuities between individual text strings. The design not only indicates to us which text strings go together and which ones do not, but also how we are to structure the cluster into a number of constituent parts. Even if it had not been for the sequential numbering of steps in this procedure, we would still be able to segment the cluster on the basis of its design (see below).

It is only in the second cluster, really, that we see the kind of classic cohesive devices, like anaphora, that we normally associate with texture in documents.

The question is: how are we to account for documents like this one in which textual relationships are realized by surface forms that are not exclusively semantic, let alone linguistic?

Perceptual cohesion

One interesting framework for discussing the structure and texture of rhetorical clusters in technical communication is Campbell's theory of perceptual cohesion (Campbell 1995). In her book *Coherence, Continuity, and Cohesion* Campbell seeks to establish some theoretical foundations for document design by providing "a set of principles for describing the unifying effects of the full range of discourse elements: from visual to semantic" (Campbell 1995: 11). In setting up this framework, Campbell draws on Grice's well-known maxims of conversation, and what is perhaps more interesting, on principles and concepts taken from Gestalt psychology.

First of all, Campbell distinguishes between local and global cohesion. *Local* cohesion may be described as surface links creating continuity between individual text elements such as words, symbols or sentences. *Global* cohesion is the realization of the *dis*continuities in discourse which enable us to perceive textual relations at a higher level, for instance between paragraphs or sections.

Our perception of continuity, and discontinuity, is, says Campbell, ultimately ascribable to the notions of *Similarity* and *Proximity*, the intensity with which these phenomena apply and the fashion in which they interact.

Local cohesion

How technical writers may design local cohesion in rhetorical clusters using similarity and proximity may be illustrated, once again, by the third cluster in figure 2, the procedure entitled "Description".

In these four steps we perceive a connection between, on the one hand, the text strings denoting the instructions and, on the other, the comments in steps 2 and 4. This perception is due to the reuse of *similar* design elements including parallel morphosyntactic structure, type fonts and bullets. When multiple similarities, linguistic or non-linguistic, interact to promote the *same* perceptual continuity, as in this case, the *Principle of Reinforcement* is said to apply.

In this example similarity also serves the purpose of signifying certain *lexical* relations. By boldfacing the words **Edit**, **Copy**, and **Paste** in steps 2 and 4, the writer signals that all three lexical items belong to the same class of entities in the world of discourse, screen elements to be precise.

The *Principle of Proximity* acknowledges that physical distance between signs may influence a reader's perception of what goes together in the text. Thus, in this fragment the difference in line spacing produces proximity-based local cohesion in steps 2 and 4 in effect making the comments (the bulleted lines) adhere, as it were, to the instructions they modify.

Take another text fragment from the same online manual (figure 3). Notice the design of the text string in step 3. Its layout and typography indicate a cohesive tie with the instructions in steps 1 and 2, even though its morphosyntactic structure and propositional content clearly signal that it is in fact a system response to the user action in step 2. What we see here is an example of an inconsistent, or ambiguous, design in which the physical appearance of a text string promotes *another* perception of continuity than its linguistic form and meaning. Using Campbell's terminology, we may say that the *Principle of Conflict* is at work here.

Figure 3

Step 1	Select the text or graphic that will serve as the link launch point (starting point) for the Program Link.
Step 2	Choose **Program Link** from the **Customize** menu. • The Link to Program dialog appears.
Step 3	The dialog lists all EXE, PIF, COM and BAT files in the current directory.

Example taken from Folio Views® Online Reference Manual

An arguably more user-friendly rhetorical clustering would probably have resulted in a design like the one represented in figure 4. A design in which the two text strings "The dialog lists all EXE, PIF COM, and BAT files in the current directory" and "The link to Program dialog appears" are designed similarly and brought closer to each other.

Figure 4

Step 2   Choose **Program Link** from the **Customize** menu.
- The Link to Program dialog appears.
- The dialog lists all EXE, PIF, COM and BAT files in the current directory.

Foregrounding

User-friendliness, or usability, may also be achieved through what Campbell calls *foregrounding*. Foregrounding can be illustrated by the visual example in figure 5. Try and locate the letter O in the two figures. Although the O is placed in the same column and in the same row in the two figures, it is considerably easier to locate the letter in the figure on the right hand side because of the stable configuration of objects against which it is perceived.

Figure 5

```
SWERTMLSPR      XXXXXXXXX
LNVCSMWQAL      XXXXXXXXX
VLCXNQRSMC      XXXXXXXXX
NLAWYETOMB      XXXXXXXOXX
PLAQMNCEW       XXXXXXXXX
```

Example adopted from Campbell (1995)

The purpose of foregrounding in *text* is to create a uniform background against which *semantic* differences may be more effectively conveyed. What this has to do with rhetorical clustering may be exemplified by the examples in figure 6, both taken from *The Microsoft Manual of Style for Technical Publications* (Microsoft 1995).

Figure 6

6a)
However, if the newly attached data file contains field names different from those inserted in the main document, you must either change the field names in the header record to match the field names in the main document or replace the field names in the main document.

6b)
However, if the newly attached data file contains field names different from those inserted in the main document, you must do one of the following:

• Change the field names in the header record to match the field names in the main document.

• Replace the field names in the main document.

Example taken from the Microsoft Manual of Style for Technical Publications (1995)

These examples show how the reformatting of a cluster may be said to make information more accessible. In example b), the difference in meaning between the two options *change the field names in the header record* and *replace the field names in the main document* are foregrounded in a text transformation involving both language and layout. In perceptual terms, one might say that the technical writer has produced a uniform background by providing the two meanings with design manifestations which are similar in terms of morphosyntactic structure, format and spatial arrangement.

Global cohesion

The way readers segment a text into clusters, or a cluster into structural parts, is bound up with their perception of *dis*continuities. Discontinuities are generated, according to Campbell, by the intensity with which *The Principles of Similarity and Proximity* apply. *The Principle of Intensity* explains, for example, why most readers would probably tend to perceive the text fragment in figure 7 as consisting of two main constituents. Here *The Principle of Similarity* operates less intensely between paragraphs one and two than between two, three and four, consequently creating greater *dis*continuity between the first two paragraphs than between the last three.

Figure 7

You can use the <DViewOnlySelect> statement to control whether active cross-references highlight the destination text. This can improve the appearance of hypertext documents.
• When cross-references are active and <DViewOnlySelect Yes> is specified, clicking a cross-reference in the document highlights the destination text.
• When cross-references are active and <DViewOnlySelect User Only> is specified, clicking a cross-reference does not highlight the destination text. The user can select text in the locked document.
• When cross-references are active and <DViewOnlySelect No> is specified, clicking a cross-reference does not highlight the destination text. The user cannot select text in the locked document.

Example taken from FrameMaker® Online Manual

The Principle of Intensity also explains why white space is so useful in document design. By applying varying amounts of white space, a text designer may enhance or reduce proximities in a text thereby suggesting where the borders of the individual text segments are to be found.

In circumstances in which *The Principles of Similarity and Proximity* apply equally intensely, another principle comes into play, namely that of *Symmetry and Size*. Figure 8 contains a text fragment consisting of six text strings surrounded by the same amount of white space. Although, in principle, a reader might structure the text fragment in a number of ways, the *Principle of Symmetry and Size* predicts that he or she will group the elements into two clusters, each consisting of three elements. The reason is that readers tend to avoid creating very small text segments and that they prefer to subdivide texts into symmetrical parts.

Figure 8

- To make cross-references emulate the openlink command, which displays the destination page in the new document, use the following statement:

```
<DViewOnlyXRef Open Behavior>
```

Use this setting to allow users to see the source page and the destination page.

- To turn off active cross-references, use the following statement:

```
<DViewOnlyXRef NotActive>
```

Use this setting to emulate the behavior in earlier FrameMaker releases.

Example taken from FrameMaker® Online Manual

The technology of text

Simply put, we may say that most digital documents today consist of two things: data (information) and markup.

Text markup is, broadly speaking, the coding, or tagging, of textual elements in a digital document. Text markup may be divided into two major types: descriptive and presentational. Descriptive markup is the tagging of structural elements of a text while presentational markup is codes specifying the physical appearance and position of text elements in the document. While presentational markup facilities exist in all decent word processing tools, descriptive markup is normally based on SGML, an international standard for coding documents across disparate platforms and products.

There are several reasons why markup is relevant in this connection. One is, as already mentioned, that markup is becoming an integral part of digital documents. One only needs to think of documents published on the World Wide Web, which must be coded in, or converted to HTML, a tagging scheme based on SGML. And there is little doubt that markup will become one of the most powerful tools for structuring, disseminating, reusing and distributing technical information products in the future.

Another reason is that (presentational) markup is in fact one of the prime vehicles for actually creating rhetorical clusters and perceptual cohesion through the interaction of text and visual design. Consider once again the procedure in figure 2. It was noted above that the difference in line spacing is instrumental in generating proximity-based cohesion. The interesting thing is, however, that the cohesion is not the result of the text designer inserting a variable number of blank lines but rather the outcome of the coding of the individual text strings. In this case the comments are hardwired to be more proximate to the preceding line than the instructions. Therefore, in a way, proximity may here be construed as a property of the individual text string just like its linguistic structure or layout.

Structure and texture in digital documents

Now, how do we relate the concepts of rhetorical cluster, perceptual cohesion and markup in digital documents with the traditional text linguistic notions of structure and texture?

Let me, tentatively, suggest a simple matrix model of digital documents on the basis of which the interrelation of these notions may be conceptualized (figure 9).

Figure 9

	Structure and texture in digital documents	
	Structure	**Texture**
Semantic level	Semantic structure (⇒ semantic structure elements)	Continuities and discontinuities
Design level	Design structure (⇒ rhetorical clusters)	Global perceptual cohesion Local perceptual cohesion
Markup level	Descriptive markup	Presentational markup

The model introduces structure and texture as two important characteristics of digital documents. It is assumed that a digital text consists of constituent parts and has texture, i.e. contains a set of mechanisms making the document, and its constituents, hang together as a unified whole. Structure and texture are relevant notions at various levels. In this model three layers are recognized:

The semantic level comprises the meaning aspects of the document. Semantic structure refers to the way the document may be divided into semantic structure elements, composite meanings with a certain overall recognizable communicative goal (definitions, warnings, procedures, etc.) and how these semantic structure elements may be further subdivided into individual messages or propositions.

Semantic texture is the continuities and discontinuities that we perceive in a text.

The design level, or surface level, contains the linguistic and/or graphic manifestations of the semantics. The design structure is the tangible, or 'visual' realization of the semantic structure. At the design level semantic structure elements may, more or less conspicuously, be represented as rhetorical clusters, which, in turn, are built up from individual discourse elements, linguistic or non-linguistic.

Thus, one may argue that rhetorical clustering is the activity, or even art or craft, of designing (appropriate) text manifestations of semantic structure elements.

As for the manifestation of texture, local and global perceptual cohesion is the stuff from which continuities and discontinuities emerge. Structurally, local cohesion may be described as the glue that binds together the individual text elements in a rhetorical cluster whereas global cohesion may be seen as the primary mechanism for keeping entire clusters apart.

To the extent rhetorical clusters in technical documentation are created as "small cohesive chunks of technical communication of predictable size, content, and appearance" (Weiss 1991: 82), we may refer to them as *modules*.

The markup level is the underlying, or invisible, layer of a digital document. Here we find information specifying the structure and/or the format and position of the data elements (descriptive versus presentational markup).

Concluding remarks

As far as genre analysis is concerned, the matrix may be used as a starting point for analysing information architectures in specific digital (sub)genres, and more specifically, the relationship between markup, design and semantics in technical (online) documentation. The model as such visualizes interfaces between relevant aspects of document design and serves to raise questions such as:

- to which extent, and how, do rhetorical clusters correlate with semantic structure elements in specific technical genres?
- how is perceptual cohesion used to create rhetorical clusters in specific technical genres?
- to which extent do rhetorical clusters reflect underlying formal structures in various forms of digital communication?

It is hoped that gaining insight into how various formal and visual design parameters interact in online documentation will generate new theories and conceptual models that perhaps more explicitly take into account that electronic genres are not only linguistic manifestations of certain communicative purposes but digital entities created, processed and distributed by computers. The need for such theories and models is felt not least in educational settings, where attempts to bridge the gap between the worlds of technology and text are increasingly reflected in curricula and syllabuses. This is especially true of technical communication programmes in which the integration of course modules such as electronic publishing and online documentation with more communicatively oriented disciplines like text linguistics and discourse analysis is seen as one step towards meeting the demands of industry for communicators who can design and produce quality (online) documentation using information and communications technology.

References

Brockmann, J.R. 1990. *Writing Better Computer User Documentation. From Paper to Hypertext.* John Wiley & Sons, Inc.

Campbell, K.S. 1995. *Coherence, Continuity, and Cohesion.* Hillsdale, New Jersey: Lawrence Erlbaum Associates, Inc.

Denton, L. and Kelly, J. 1993. *Designing, Writing and Producing Computer Documentation.* McGraw-Hill.

Folio Views® Online Reference Manual 1994. Folio Corporation. Provo, USA.

FrameMaker® Online Manual 1995. Frame Technology Corporation. San Jose, USA.

Halliday, M.A.K. and Hasan, R. 1985. *Language, Context and Text: Aspects of Language in a Social-Semiotic Perspective.* Victoria: Deakin University Press.

Horn, R. E. 1989. *Mapping Hypertext.* The Lexington Institute.

The Microsoft Manual of Style for Technical Publications. 1995. Microsoft Press.

Moore, P. and Fitz, C. 1993. Using Gestalt Theory to Teach Document Design and Graphics. *Technical Communication Quarterly,* Vol. 2, No. 4, 389-410.

Schriver, K.A. 1997. *Dynamics in Document Design.* John Wiley & Sons, Inc.

Turner, R.C., Douglass, T.A. and Turner, A.J. 1996. *Readme.1st. SGML for Writers and Editors.* Upper Saddle River, New Jersey: Prentice Hall PTR.

Weiss, E.H. 1991. *How To Write Usable User Documentation.* Oryx Press.

Applied Genre Analysis

The Role of Genre for Translation

CHRISTINA SCHÄFFNER
Aston University, Birmingham

Introduction

In the preface to a book on Academic Writing, the editors Eija Ventola and Anna Mauranen state: "We still do not know very much about the linguistic and textual features which characterize successful products and distinguish them from unsuccessful ones." (1996: vii). This statement can be applied in a similar way to translation, i.e. how do we know that a target text (TT) is a "good" one? one of high quality? The answers to this question will be different, depending on underlying assumptions about the nature of translation. For scholars working within a linguistic approach to translation, a good translation will be one which reproduces the meaning of the source text (ST) as accurately and closely as possible. This is the position of Newmark, for example, who requires the TT to be accurate "in denotation and in connotation, referentially and pragmatically" (Newmark 1991: 111). Textlinguistic approaches have led to an increased awareness of the fact that the linguistic format of the TT is above all determined by target language text-typological conventions as well as by aspects of the communicative situation in the target culture in which it is to fulfil its function (e.g. time and place, knowledge and expectations of the TT addressees). Functionalist approaches to translation (e.g. Skopos theory, the theory of translatorial action, cf. Reiß and Vermeer 1991, Hönig and Kußmaul 1992, Nord 1993, Holz-Mänttäri 1984) focus on the purpose of the TT as the decisive criterion for any evaluation of translation quality (cf. Schäffner 1998).

Translation is above all text production. In other words, the aim of any translation is to produce a target text that fulfils its specified purpose. A 'good' translation is thus no longer a correct rendering of the ST, in the sense of reproducing the ST meanings of micro-level units. It is rather a TT which effectively fulfils its intended function in the target culture. One of these functions may of course be the exact reproduction of the ST, but this is not the only one. Instead of 'good', some translation scholars prefer to speak of '(pragmatically) adequate' or of 'functionally appropriate' translations. In such a prospective view of translation, the structure and the actual linguistic make-up of the TT are determined by various factors, especially by the intended purpose and function of the TT, by text-typological and/or genre conventions, the addressees' background knowledge and their communicative needs, as well as by the ST. The relevance and the weight of these factors differ in each specific situation.

In this paper, I will concentrate on one of these factors, namely text-typological and/or genre conventions. After a short discussion of the concepts of text type and genre in textlinguistic research, I will illustrate how they have been applied in translation studies. Finally, some pedagogical implications for translator training will be discussed.

Text type - genre - Textsorte

One of the aims of textlinguistic research has been the classification of texts. Various criteria, both text-external and text-internal ones, have been used for arriving at a typology of texts. Some typologies are based on a dominant communicative function, or the communicative purpose, of the text (e.g. Isenberg 1984, Werlich 1975). In such a perspective, the scholars have usually set up a limited number of categories. For example, Werlich's typology has five idealised text types (description, narration, exposition, argumentation, instruction).

These basic text types are then linked to specific genres. There is, however, sometimes some terminological confusion in that 'text type' and 'genre' are used synonymously (cf. also Trosborg 1997). In German textlinguistic literature, there is normally a differentiation between 'Texttyp' and 'Textsorte' with a different theoretical basis for their categorisation. 'Texttyp' (text type) is understood as a category for a more abstract, theoretical classification of texts, whereas 'Textsorte' (or 'Textklasse', i.e. genre, text class) is a label used for an empirical classification of texts as they exist in a human society (cf. Heinemann and Viehweger 1991: 144). Genres (Textsorten) are defined as global linguistic patterns which have historically developed in a linguistic community for fulfilling specific communicative tasks in specific situations. They are a kind of generalisations, based on the experience of the speakers of a communicative community. Genres reflect the effective, conscious and situationally appropriate choice of linguistic means. Members of a linguistic community can be said to have a specific genre knowledge, i.e. a knowledge of global text patterns, rather than a text type knowledge (cf. Heinemann and Viehweger 1991: 144).

German-speaking authors, when writing in English, sometimes use 'text type' as equivalent to 'Textsorte' (maybe because 'Genre' in German is restricted to literary genres). For example, Suter's definition of 'text type' corresponds to what is typically called 'genre' in Anglo-Saxon studies on genre analysis. This becomes obvious when we compare the definitions by Suter and Swales:

> "A traditional text type is what a given speech community, at a given time and over a considerable period of time, accepts as a traditional, conventional and in some specific way linguistically standardised textual model to be constantly re-used for specific communicative purposes." (Suter 1993: 48).

> "A genre comprises a class of communicative events, the members of which share some set of communicative purposes. These purposes are recognized by the expert members of the parent discourse community, and thereby constitute the rationale for the genre. This rationale shapes the schematic structure of the discourse and influences and constrains choice of content and style. ... In addition to purpose, exemplars of a genre exhibit various patterns of similarity in terms of structure, style, content and intended audience. ..." (Swales 1990: 58)

Instead of setting up text types, textlinguists have recently been more concerned with classifying, defining and describing genres. The classification of genres is again often based on communication functions, thus showing some overlap with the categorisation of text types; for example, Rolf's five groups of genres, i.e. assertive, directive, commissive, expressive, declarative, which he further specifies into subgroups (Rolf 1993).

Genres are embedded in sociologically-determined communicative activities. They can be described as conventional, typical combinations of contextual (situational), communicative-functional, and structural (grammatical and thematic) features. In other words, in analysing genres, their structures, both at macro- and micro-level, are systematically related to the communicative function these genres (are meant to) fulfil. Some genres have been found to be highly conventionalized (e.g. weather reports, recipes, contracts), whereas others conform to a more flexible set of conventions (e.g. advertisements, poems, novels). For some genres, these conventions seem to be highly culture-specific, i.e. different from culture to culture, whereas for others, the conventions seem to be more or less universal, or at least supra-cultural. But even in the case of supra-cultural or universal conventions, the specific linguistic realisation in the individual languages and cultures (or: languacultures, cf. Agar 1992) may differ. For example, it can be said that instruction manuals share the communicative function of 'giving instructions'. However, at the textual micro-level, whenever the instructions are arranged in a sequence to indicate the individual steps of the action, this function is conventionally realised by imperatives in English (e.g., 'Switch on the radio'), and by infinitives in German (e.g. 'Gerät einschalten', cf. Kußmaul 1990).

This culture specificity of genres, as well as their normative effect (i.e., due to their conventional structure, genres provide some orientation for the production and reception of texts), makes them interesting and relevant for translation studies too. In the same way as users of a specific language recognize a text as an instance of a particular genre, translators will recognize the ST as an instance of a genre. But even more: as experts in interlingual and intercultural communication, translators will also have an active competence in a wide variety of genres, combined with a textual expertise both in the source and in the target language, i.e. they know how to produce the TT as an instance of the required genre for the target culture.

Genres in translation studies

Also in translation studies, the terms 'genre' and 'text type' are often used interchangeably, as can be seen in the similarity of the definition of genres as "conventionalized forms of language use appropriate to given domains of social activity and to the purposes of participants in them" (Hatim and Mason 1997: 39), and of text types as "sets of socially situated textual instances which share certain features" (Neubert and Shreve 1992: 127).

One of the first translation scholars to point out the importance of a categorisation of texts for translation purposes was Katharina Reiß (1971), although she was mainly concerned with text types and not with genres. The aim of her translation-oriented text typology was to derive strictly objective criteria for assessing the quality of translations. Based on Bühler's three functions of language, 'Darstellung' (description, presentation), 'Ausdruck' (expression), and 'Appell' (appeal), Reiß derived three corresponding dimensions of language (logisch, ästhetisch, dialogisch, i.e. logical, aesthetic, dialogical) and three corresponding text types (inhaltsbetont, formbetont, appellbetont, i.e. informative, expressive, appellative; she also added an audio-medial text type as a fourth type, which will not be further considered here).

These three text types are then linked to translation methods. According to Reiß, for the informative text type (e.g. report, textbook, instruction manual) the aim is invariance of content; the translation method is therefore a straightforward one, with the focus on the content of the text and on transmitting the information in full. In the case of the expressive text type (e.g. novel, poem, comedy), where we have an artistic shaping of the context, the aim is an analogy of the artistic form, and the translation method is called identifying. For the operative text type (e.g. advertising, church service, propaganda leaflet), the aim is identity of the text-internal appeal, production of identical behavioural reactions; and the translation method is adaptation. Based on her text types and translation methods, Reiß provides guidelines for how to deal with specific translation problems. She argues, for example, that a metaphor or an idiom in an expressive text must be rendered as a metaphor or an idiom in the target text, whereas this is not necessary for a metaphor or an idiom in an informative text.

Reiß's text types ('Texttyp'in German) can be realized by different text varieties ('Textsorte' in German). For example, a private letter would belong to the informative text type, whereas a begging letter would belong to the operative text type. However, Reiß does not devote much space to translation problems arising from these varieties, it is rather the text types which she relates to translation methods.

Reiß's text typology has often been criticized as too rigid, and her translation methods as too prescriptive. Snell-Hornby (1988: 31) commented that what is wrong with Reiß's typology "is the use of box-like categories as a kind of prescriptive grid, creating the illusion of clear-cut objectivity." In her own study, Snell-Hornby applies prototype theory to a text typology to demonstrate that text types do not display clear-cut features, but that there are blurred edges and overlappings. In a diagram she illustrates a system of relationships between basic text types - as prototypes - and the crucial aspects of translation (cf. Snell-Hornby 1988: 32). On the vertical plane, her diagram represents a stratificational model of six levels which proceeds from the most general level (A), presenting the conventional areas of translation (literary, general language, special language translation), downwards to the most particular level (F), listing phonological aspects of specific relevance for certain texts, e.g. alliteration in advertising.

Snell-Hornby's level B presents a prototypology of the basic text types of which only a selection are given: Bible, stage/film, lyric, poetry, modern literature, classical antiquity, literature before 1900, children's literature, light fiction (related to literary translation on level A); newspaper/general information texts, advertising language (related to general language translation); legal language, economic language, medicine, science/technology (related to special language translation). However, the way these text types are labelled gives the impression, that what we actually get is a mixture of subject areas and genres.

Instead of listing prescriptive translation methods, Snell-Hornby points out important aspects and criteria which are of relevance in the translation process. For example, she lists some of the following hypotheses:

"the more 'specialized' or 'pragmatic' the source text, the more closely it is bound to a single, specific situation, and the easier it is to define the function of its translation;

the more specific the situation and the more clearly defined the function, the more target-oriented a translation is likely to be; ...

the more 'literary' a translation, the higher is the status of the source text as a work of art using the medium of language" (Snell-Hornby 1988: 115).

Translation as TT-production means - for a large number of texts, and if the skopos does not demand otherwise - adopting the target text to the genre conventions of the target culture, i.e. conventions the TT addressees are familiar with and would expect in the specific situation. As noted above, knowledge about (culture-specific) genre conventions is therefore of relevance for producing appropriate target texts. Knowledge of cross-cultural similarities and/or differences regarding genres and genre conventions is crucial to the translator.

Scholars working within a textlinguistic approach to translation have repeatedly stressed the importance of uncovering the typical conventions of texts (at macro- and micro-levels) and, thus, of discovering text type specific translation regularities, whether they prefer the term genre (e.g. Hatim and Mason 1997), text type (e.g. Neubert 1985, Kußmaul 1995), or Textsorte (e.g. Göpferich 1995). They advocate the construction of prototypes of text types (e.g. Neubert and Shreve 1992, Kußmaul 1995), or of text type profiles ('Textsortenprofile', Göpferich 1995) as a result of corpus-based contrastive studies. On the basis of such studies it should be possible to determine how specific conventions are expressed in the source and target languages.

Knowledge about particular genres can be gained from a systematic analysis of parallel texts, which are "L_2 and L_1 texts of equal informativity which have been produced in more or less identical communicative situations" (Neubert 1985: 75). Parallel texts are useful for assessing how the same kind of factual material is verbalized in different languages, how identical communicative functions are expressed in specific genres of SL and TL, and, thus, consequently, for providing translators with guidelines for producing a TT which corresponds to the target culture conventions. In other words, genre prototypes, or genre profiles, can function as models for the translator.

In recent years, an increasing number of studies have investigated the characteristic features of genres, both a macro- and micro-level, from a translational perspective. Examples of such translation specific comparative genre analysis are the following: Arntz (1992) revealed different macrostructures in American, French, German, Italian and Spanish legal texts; Kußmaul (1990) systematically related the infinitives in German and the imperatives in English instruction manuals to the function of the macrostructure category in which they are predominantly used; Lawson (1983) discovered that English, French, and German patents are fairly identical in their rigid macrostructures, due to international norms, but that they are different in their microstructures; Hatim and Mason (1997) illustrate cultural differences in the argumentative style in English and Arabic, which have a preference for counter-argumentation and through-argumentation, respectively. Snell-Hornby (1984) revealed significant differences in the linguistic form of typical speech acts (request, command, warning, prohibition) in English and German public notices. For example, in English public notices, imperatives and a direct form of address are typically used, together with an identification of the addressees (in their situational roles) as agents (agentive nouns). A variety of grammatical structures are used to express prohibitions (e.g., modal verbs, determiner 'no'). In German public notices, infinitives are preferred to express these speech acts, as are abstract nouns (verbal nouns, focusing on the action). To express prohibitions, the preferred forms are the past participle 'verboten', or its more formal stylistic variants 'untersagt' and 'nicht gestattet'. These differences are illustrated in the following examples: 'Hawkers, canvassers, collectors not allowed' compared to 'Hausieren verboten'; 'No smoking' compared to 'Rauchen verboten'.

In an excellent and comprehensive study with immediate relevance for translation studies, Göpferich (1995) analysed a number of German and English genres (e.g. instruction manuals, patent specifications, conference reports) for their conventional macro- and microstructures. She found for example, that conference reports and book reviews seem to have more flexible macrostructures, in contrast to patent specifications in the scientific and technical areas. Moreover, German and English patents are fairly similar, due to the fact that science and technology are international fields, and therefore such texts are normally not significantly influenced by their respective individual cultures.

Using genres in translation teaching

The relevance of genres and genre conventions has been widely recognized in translation studies, especially in a textlinguistic approach to translation. Working with genres is equally useful in translation teaching, since such an approach helps to increase students' sensitivity to the linguistic patterning at various levels and thus makes them aware of the complexity of translation competence. For pedagogical purposes it is worthwhile to use both genres that show little or hardly any culture-bound differentiation in their structures and genres that are indeed firmly embedded in individual cultures. A translation course which is build on a selection of genres has certain advantages, among others, that students can be asked to include parallel texts in the preparation of their translation assignments (cf. Schäffner in press). (More or less conventionalized) genres that we have used in English-German and German-English translation courses include instruction manuals, tourist brochures, contracts, film and book reviews, job offers, news reports. The students are asked to find exemplars of the genre in question, both in the source and in the target language, and then look for systematic regularities for the genre in both languages and cultures. Based on such comparisons they come up with their TT versions. Corresponding structures available for use in the TT are thus taken from actual usage in parallel texts. In other words, parallel texts function as reference tools and as instruments to identify the conventions associated with particular genres.

The advantage of parallel text analysis will be illustrated here on the basis of two sample genres, job offers and news reports. With increasing internationalisation, as for example in the European Union, it becomes more and more common for companies also to advertise vacant positions outside their home culture. Although the dominant function of this genre is the informative function, the texts contain an element of instruction, typically at the end of the text where interested candidates are told where to send their application forms and by which deadline. This speech act is expressed differently in the English and German cultures, although both make use of auxiliary verbs. Students were asked to compare original English and German job offers for university positions, and on the basis of this parallel text analysis to find the typical, recurring structures that are used to express the instructions. Typical English formulations are:

applications should be sent to [address]
applications should be submitted by [date]
applications should be lodged not later than [date]
applications should be received not later than [date]
application forms to be returned by [date]
application forms (returnable by [date]) are available from [address]
for an application package, please contact [address]
for further details telephone

Typical German formulations are:

Bewerbungen sind bis [date] zu richten an [address]
Bewerbungen sind an [address] zu senden
Bewerbungen werden erbeten an [address]

Bewerbungen richten Sie bitte an [address]
Richten Sie bitte Ihre Bewerbung bis [date] an [address]

Forms with the modal verb 'should' plus passive (should be sent) could be identified as the most frequent ones in the English job offers, whereas in the German ones, the formulation with 'sein' plus infinitive (sind zu richten, literally: are to be sent) and the polite imperative (richten Sie bitte) seem to be equally frequent. These findings confirm Kußmaul's observation that in English, the imperative, which may be considered as the prototypical form for giving instructions, may be used in a large number of situations and genres but not in regulations and rules, as e.g. in job offers (cf. Kußmaul 1997: 77).

Job offers are increasingly translated. An analysis of some offers from German universities and academic foundations written in English (whether these are translations or not is not absolutely clear) revealed interesting results. Some were formulated following the patterns of the English texts, i.e. using a form like 'applications should be sent' for instructions, or, to give two more examples of formulations that had been identified as typical of English job offers, 'the successful applicant will have a PhD' (in the German texts, we frequently find a heading 'Einstellungsvoraussetzungen', followed by a list of expected qualifications and skills), and 'The University XYZ is an equal opportunities employer' (in German job offers, a very frequent formulation is 'Die Universität XYZ fordert Frauen ausdrücklich zur Bewerbung auf. Schwerbehinderte werden bei gleicher Qualifikation bevorzugt berücksichtigt.'). In the case of some other job offers, the English structure mirrored very much that of the German text, for example in the instruction 'applications are to be submitted to [address]', or in a more literal translation of the passage referring to equal opportunities as 'The XYZ University strives to promote women in teaching and research and therefore particularly encourages this group to apply. The university gives preference to disabled people in cases of identical personal and professional suitability.'

The advantages of a translation-oriented text comparison are most obvious in the case of highly conventionalized genres. Genre conventions can be said to lead to intercultural translation problems, to use Nord's typology of translation problems (e.g. Nord 1991, 1997). They arise from the differences in the conventions between the two cultures involved, including genre conventions. My second example, a journalistic text, or more precisely, a news report, does not represent a highly conventionalized genre. However, a parallel text analysis proved extremely useful in an introductory translation course with first-year students. They were given an English newspaper text on UK car exports, reporting on developments as to the number of exports compared to imports, and quoting the expectations of some economists. The students were asked to analyse German (and French) texts to find the specific conventions in these target cultures for five characteristic features that had been identified in a pre-translational ST analysis. These five features of our journalistic text, i.e. the specifically identified translation problems, and the corresponding German structures, that are meant as guidelines for translating this text, are given below:

(1) references to the countries of origin of some particular industry or company

English ST: 'the UK car industry'

German: adjective, with exception of (sometimes) 'US' (e.g. 'der schwedische Nutzfahrzeughersteller X, der US-Autokonzern X')

(2) references to culture-specific institutions of other countries

English ST: 'Society of Motor Manufacturers and Traders, ... SMMT'

German: often use of adjectives for countries, a general word for the type of institution (e.g. 'Handelsorganisation', 'Landwirtschaftsverband'); proper name of institution either translated, but more often kept in original; abbreviation always in the original form, once introduced together with full name, the following text will use only the abbreviation; e.g.:

Die holländische Handelsorganisation *Vieh, Fleisch und Eier (PVE)* ...

Die Singapore Airlines (SIA) ...

Der schwedische Landwirtschaftsverband (LRF) ...

(3) references to people, their names and positions, titles

English ST: 'M. P., SMMT chief economist'

German: Two possibilities: either full name followed by title/position (this seems to be the preferred strategy when title/position is relatively long), or title/position followed by full name (preferred when title/position is just one word, or when position is rheme and more stressed); in case of women, the female form for the title/position is used; sometimes a more general form ('für ... zuständig') when the exact position is either irrelevant or not known; e.g.:

... B. R., Geschäftsführer von Holiday Inn Deutschland

... G. F., bei der brasilianischen Zentralbank für internationale Angelegenheiten zuständig, ...

Finanzdirektor H. C. ...

der französische Finanzminister J. A. ...

... die Geschäftsführerin der neugegründeten Deutschland Tourismus Marketing GmbH (DTM), U. S. ...

(4) indirect speech
English ST: 'M. P., SMMT chief economist, said export growth had offset ...'
German: In the case of indirect speech, the verbs are typically in the subjunctive (normally present subjunctive, sometimes past subjunctive). The subjunctive form indicates that there is still a reference to indirect speech, even if this is not marked by a verb of saying. For the English verb 'say', there is more variation in German texts (e.g. 'sagen, erklären, mitteilen, berichten'), especially when the text repeatedly refers to statements of a speaker, as in the last example below. When only part of a statement is quoted as direct speech, but the verb is not part of this direct speech, then again the verb is put in the subjunctive (cf. the third example below); e.g.:

... erklärte das Unternehmen
Mit einem Rückgang von 4,8 Prozent auf 29,73 Millionen Übernachtungen habe die Branche 1996 das schlechteste Jahr seit 1965 erlebt, teilte das Bundesamt für Statistik, Bern, mit.
Die Fluglinie sei "...unter Druck", sagte Vizepräsident C. K..
Das sagt I. B., Volkswirt bei der X Bank in London, voraus. ...Der Handel ohne Schranken führe, wie B. erklärte, zu einer Kostenentlastung für die Unternehmen. Die enge Verflechtung der Finanzmärkte sorge dafür, daß ..., fügte B. hinzu. ... Eine solche Regierung, meint B., werde einer Teilnahme an der Europäischen Währungsunion positiv gegenüberstehen.

(5) figures and numbers in relation to currency
English ST: '£5.5.billion'
German: Foreign currencies are most frequently converted into DM, and both values are given, the equivalent German sum either in brackets, or immediately after the original sum, or indicated by the addition of 'umgerechnet'; sometimes, the approximate German sum is indicated by 'etwa', or 'rund'. More rarely, only the equivalent German sum is given (as in the third example below), or only the original sum/currency; e.g.:

1,5 Milliarden Franc (440 Millionen DM)
2,71 Milliarden schwedische Kronen (etwa 620 Millionen DM)
könnte sich der neuerliche Finanzbedarf des X auf umgerechnet bis zu 9 Milliarden DM belaufen
1,545 Milliarden Pfund, umgerechnet etwa 4,2 Milliarden DM ...

Based on such an analysis of samples of target language parallel texts, a more objective decision as to the choice of the functional equivalents (i.e. contextually and situationally adequate lexical and syntactic structures) for the target text can be made. The aim of such an approach, i.e. the pedagogical implication, is to free the students of their (often encountered) ST-fixation which would frequently result in a too literal translation, and instead give them confidence and evidence on which they can base their translation decisions.

However, not all genres are equally influenced by conventional structures, which also puts limits to a translation-oriented parallel text analysis. Different genres display different characteristic features, and thus, pose different problems for translation. Moreover, also in the case of conventionalized genres, patterns and elements are not always equally predictable. In a recent book, Wilss argues as follows:

> "The classification of texts with a view to establishing a correlation between text and text type [Wilss uses 'text type' in the sense of 'genre' - C.S.] makes sense, albeit in a somewhat idealized fashion. However, it seems that, at least for the time being, text-oriented TS [= translation studies - C.S.], besides its attempt to discover text type-specific translation regularities, must be aware that a large portion of texts contains an 'episodic' element with stylistically more or less marked options. ... The occurrence side by side of (rhetorically) obligatory and (stylistically) optional text elements varies from text type to text type, and the translator has to proceed accordingly." (Wilss 1996: 21).

One genre which is characterized by a combination of obligatory and optional elements is academic writing. It has been influenced by different intellectual styles (cf. Galtung's (1985) saxonic, teutonic, gallic and nipponic intellectual styles). Comparing examples of English and German, Clyne (1991) has found out that English academic texts are marked by linearity, few footnotes, more personal constructions, more metacommunicative utterances and bridging sentences, with quotations and references usually integrated within the main text - in general, they are more oriented towards the reader. For German academic writing, on the other hand, the texts are marked by syntactic complexity, a large number of nominalizations, impersonal constructions, agentless passives, various modal impressions (often hedgings), a rather high number of digressions (e.g. more detailed references to some theory, a historical overview), quite a lot of footnotes - in general, they are more oriented towards content. Based on Clyne's studies, Cmejrkova (1996) looked at the Czech academic register and discovered that it inherited a lot from the German intellectual style, both its syntax and terminology and also the general ideas on the purpose of academic writing.

The reason for the different design of the texts are the completely different conceptions of academic writing. As Clyne (1991: 65) argues:

> "Knowledge is idealized in the German tradition. Thus, texts are written to transmit knowledge, and the onus is on the reader to make the effort to under-

stand them in order to benefit from this knowledge. German texts can afford to be less easy to read. In English-speaking countries, most of the responsibility falls on writers to make their texts readable."

That the Czech tradition has been influenced by this German conception is confirmed by Cmejrkova (1996: 144): "This presupposition that it is the reader's responsibility to understand rather than the writer's responsibility to write it understandably also seems to be deeply rooted in the Czech stylistic tradition."

What then, are the problems academic writing poses for translation? For example, Kußmaul (1997: 72) asks: "Should we preserve the cultural macrostructures or are we allowed to change them? ... when translating English academic texts into German, do we have to insert digressions and footnotes?" And later on: "Should translators change the style of academic writing?" (Kußmaul 1997: 81). In referring to the difference of impersonal constructions in German texts but personal formulations in English texts, Gutknecht and Rölle (1996: 256) give a rather prescriptive advice: "In those cases where one realizes that the English speaker, even though using tentative-subjective formulations, presents objective facts, one should (sic) employ German constructions conveying objectivity. This principle would apply to both colloquial and scientific English." Kußmaul's argument, which is based on a functionalist approach (in contrast to Gutknecht and Rölle's linguistic approach), is more cautious and based on the complexity of any translational activity. His argument, "I think readability is the most important quality of a text in any culture, and all other considerations should be subject to this aim" (Kußmaul 1997: 72), is motivated by focusing on the purpose of the target text as the dominant criterion for translation. And when he says "As translators we will have to weigh the status of the source-text authors, our own status and the status of the readership carefully against each other when making these kinds of decisions" (Kußmaul 1997: 81), he brings in the aspect of translation ethics.

Clyne's 1991 paper is entitled 'The dilemma of the German-speaking scholar'. With using 'dilemma' he wants to stress that a German-speaking scholar is faced with a difficult choice: when writing in English, he or she either follows the norms and conventions of the Anglo-Saxon (i.e. specifically American) culture which dominates the market (i.e. most journals are published in English), or he or she sticks to the German conventions, thus risking that the paper will be rejected by the editors. The same would apply to a situation where a German-speaking scholar has an article translated into English. In this case, Clyne's title can equally be extended to the translator who also faces a dilemma. Should a translator produce a TT in such a way that it corresponds to the Anglo-Saxon conventions of academic writing, thus obeying the rules of the market? Or should s/he stick to the surface structure of the German ST, thus (wrongly?)

respecting the authority of the scholar? In his article, Clyne does not give a clear answer to his question, and I, too, would rather avoid any clear answer as far as the translator is concerned. But at least it should be said that it is an essential part of a professional code of ethics for the translator to contact the scholar (i.e. the client) and to point out that there are different styles in academic writing and come to some mutual agreement (cf. Kußmaul's argument concerning the status above).

To conclude: Based on a translation-oriented contrastive analysis of genres, i.e. both highly conventionalized and less conventionalized ones, it will become possible to formulate guidelines and recommendations as to the most appropriate structures and formulations of the target text - depending, of course, on the particular translation assignment, the skopos. However, such a genre analysis must not be static. Genre conventions are determined by culture and, thus, prone to constant change. In other words, a dynamic perspective needs to be applied to genre analysis. The relevance of a translation-specific (comparative) genre analysis has been recognised in translation studies. Some studies are already available, but still a lot needs to be done before we have a solid knowledge of the "linguistic and textual features which characterize successful products", both in a source and a target language and culture.

References

Agar, M. 1992. "The Intercultural Frame". unpublished ms.
Arntz, R. 1992. "Interlingualer Fachsprachenvergleich und Übersetzen." In M. Snell-Hornby, F. Pöchhacker, and K. Kaindl (eds), *Translation Studies. An Interdiscipline*. Amsterdam: Benjamins, 235-246.
Clyne, M. 1991. "The sociocultural dimension: The dilemma of the German-speaking scholar". In H. Schröder (ed), *Subject-oriented texts*. Berlin and New York: de Gruyter, 49-67.
Cmejrková, S. 1996. "Academic Writing in Czech and English". In E. Ventola and A. Mauranen (eds), *Academic Writing. Intercultural and textual issues* (Pragmatics & Beyond New Series 41). Amsterdam: Benjamins, 137-152.
Galtung, J. 1985. "Struktur, Kultur und intellektueller Stil". In A. Wierlacher (ed), *Das Fremde und das Eigene*. München: Judicium, 151-193.
Göpferich, S. 1995. *Textsorten in Naturwissenschaft und Technik. Pragmatische Typologie - Kontrastierung - Translation*. Tübingen: Narr.
Gutknecht, Ch. and Rölle, L. J. 1996. *Translating by Factors*. New York: State University of New York Press.
Hatim, B. and Mason, I. 1997. *The Translator as Communicator*. London: Routledge.
Heinemann, W. and Viehweger, D. 1991. *Textlinguistik. Eine Einführung* (Germanistische Linguistik 115). Tübingen: Niemeyer.

Hönig, H. and Kußmaul, P. 1982. *Strategie der Übersetzung. Ein Lehr- und Arbeitsbuch.* Tübingen: Narr. (4th edition 1991).

Holz-Mänttäri, J. 1984. *Translatorisches Handeln. Theorie und Methode.* Helsinki: Suomalainen Tiedeakatemia.

Isenberg, H. 1984. "Texttypen als Interaktionstypen". *Zeitschrift für Germanistik* 5(2): 261-270.

Kußmaul, P. 1990. "Instruktionen in deutschen und englischen Bedienungsanleitungen". In R. Arntz and G. Thome (eds), *Übersetzungswissenschaft. Ergebnisse und Perspektiven.* Tübingen: Narr, 369-379.

Kußmaul, P. 1995. *Training the Translator.* Amsterdam and Philadelphia: Benjamins.

Kußmaul, P. 1997. "Text-Type Conventions and Translating: Some Methodological Issues". In A. Trosborg (ed), *Text Typology and Translation.* Amsterdam/Philadelphia: Benjamins, 67-83.

Lawson, V. 1983. "The Language of Patents. A Typology of Patents, with Particular Reference to Machine Translation." *Lebende Sprachen* 28: 58-61.

Neubert, A. 1985. *Text and Translation* (Übersetzungswissenschaftliche Beiträge 8). Leipzig: Enzyklopädie.

Neubert, A. and Shreve, G. M. 1992. *Translation as Text.* Kent and London: Kent State University Press.

Newmark, P. 1991. *About Translation.* Clevedon: Multilingual Matters.

Nord, C. 1991. *Text Analysis in Translation.* Amsterdam: Rodopi.

Nord, C. 1993. *Einführung in das funktionale Übersetzen.* Tübingen: UTB.

Nord, C. 1997. *Translating as a Purposeful Activity. Functionalist Approaches Explained.* Manchester: St. Jermone.

Reiß, K. 1971. *Möglichkeiten und Grenzen der Übersetzungskritik.* München: Hueber.

Reiß, K. and Vermeer, H. J. 1991. *Grundlegung einer allgemeinen Translationstheorie*, 2nd edition, (= Linguistische Arbeiten 147). Tübingen: Niemeyer.

Rolf, E. 1993. *Die Funktionen der Gebrauchstextsorten.* Berlin: de Gruyter.

Schäffner, C. in press. "Parallel texts in translation". in M. Cronin (ed), *Translation Studies - Unity in Diversity?* Manchester: St Jerome.

Schäffner, C. (ed.). 1998. *Translation and Quality.* Clevedon: Multilingual Matters.

Snell-Hornby, M. 1984. "The linguistic structure of public directives in German and English". *Multilingua* 4: 203-211.

Snell-Hornby, M. 1988. *Translation Studies. An Integrated Approach.* Amsterdam/Philadelphia: Benjamins.

Suter, H.-J. 1993. *The Wedding Report: A Prototypical Approach to the Study of Traditional Text Types.* Amsterdam/Philadelphia: Benjamins.

Swales, J. 1990. *Genre Analysis. English in academic and research settings.* Cambridge: Cambridge University Press.

Trosborg, A. 1997. "Text Typology: Register, Genre and Text Type". In A. Trosborg (ed), *Text Typology and Translation.* Amsterdam/Philadelphia: Benjamins, 3-23.

Ventola, E. and Mauranen, A. (eds). 1996. *Academic Writing. Intercultural and textual issues* (Pragmatics & Beyond New Series 41). Amsterdam: Benjamins.

Werlich, E. 1975. *Typologie der Texte*. Heidelberg.
Wilss, W. 1996. *Knowledge and skills in translator behavior*. Amsterdam/Philadelphia: Benjamins.

Broadening the Perspective

Analysing LSP Genres (Text Types): From Perpetuation to Optimization in Text(-type) Linguistics[1]

SUSANNE GÖPFERICH

Karlsruhe

In the last 15 years, numerous LSP genres have been analyzed and compared intralingually as well as interlingually and interculturally. The results of these analyses, however, have hardly been related to each other and consequently not been joined together to form a mosaic that gives us an overview over larger text-type systems. The methods of such an integration process will be the subject of the first part of this article. In the second part, I will raise the question whether the characteristics of specific genres which we find out on a purely descriptive basis actually contribute to the fulfilment of these genres' communicative functions in an ideal way. Furthermore, I will discuss methods which may help us to answer this question and, thus, to develop guidelines for technical writers.

The present situation in LSP text linguistics

For the last 15 years, LSP research has focused on LSP texts, especially LSP genres or text types. Previously, LSP researchers had mainly been concerned with terminology and syntax. Kalverkämper's paper "Textuelle Fachsprachen-Linguistik als Aufgabe" ("Textual LSP Linguistics as a Task") (1983) marks the transition from lexicology-oriented to text-oriented LSP research: "Texts as they actually occur, i.e., as real parole phenomena (individual texts), as well as texts as abstract entities, i.e., as entities in the langue system, must be the sole and decisive points of departure for language descriptions and consequently also LSP descriptions." (Kalverkämper 1983: 126; my translation)

Since the publication of Kalverkämper's paper, the necessity of integrating text (and pragma-) linguistics into LSP research has been voiced again and again. In 1985 Baumann (1985: 142), for example, considered the close examination and exact differentiation of LSP genres one of the most urgent tasks facing LSP research at that time. Hoffmann, too, has emphasized the importance of LSP genre analyses:

> "An exact classification of scientific and technical texts on the basis of dominant distinguishing features forms one of the decisive prerequisites for the successful solution of specific communicative tasks [...]. A catalogue and later a systematic description of the different types of texts which are mainly determined by their specific functions, together with their typical features may contribute to improved comprehension of specialized information and – above all – to the production of texts which convey such information adequately. This, again, would be a major contribution to more efficient mono- and multi-lingual communication." (Hoffmann 1983: 62; my translation)

Here Hoffmann appears to assume that the features which prove to be characteristic of a particular genre also contribute to the fulfilment of its communicative function in an ideal way and should thus be imitated. I will come back to this assumption.

Hoffmann is not the only one to emphasize the usefulness of genre analyses – especially contrastive ones. Fluck (1991: 218 ff.), Kußmaul (1978: 57) and Reiß and Vermeer (1984: 194), to name only a few, have also stressed their importance. In 1991 "Comparison as a Method in LSP Research" ("Der Vergleich als Methode in der Fachsprachenforschung") was chosen as the motto of the first LSP conference which brought researchers from East and West Germany together in Leipzig (October 17 and 18, 1991). The conference results were published in a volume entitled *Kontrastive Fachsprachenforschung (Contrastive LSP Research)* (Baumann and Kalverkämper 1992), which again focuses on comparisons of LSP genres. This is no wonder since LSP research and LSP genre analyses had almost become synonymous by this time (cf. Baumann and Kalverkämper 1992: 24).

In the last 15 years, the central role of LSP texts and LSP genres in LSP research has resulted in numerous genre descriptions as well as intralingual and interlingual comparisons of genres. This work culminated in the volume *Fachliche Textsorten: Komponenten – Relationen – Strategien (LSP Genres: Components – Relations – Strategies)* edited by Kalverkämper and Baumann (1996), which was dedicated to Lothar Hoffmann, the "nestor of LSP research" (Baumann and Kalverkämper 1996: 16). Apart from a few exceptions, however, the genre analyses carried out so far have been limited to specific genres such as patent specifications (Raible 1972, Schamlu 1985), papers in the humanities (Kußmaul 1978), English chemistry articles (Weise 1980), weather reports (Spillner 1983), articles in medical journals and their popularized versions (Adams Smith 1987), operating instructions (Kußmaul 1990), abstracts (Kretzenbacher 1990, Adolphi 1996), introductions to scholarly papers (Gnutzmann and Lange 1990), conclusions (Oldenburg 1992), lexicon articles (Schaeder 1996), annual business reports (Bolten et al. 1996), prefaces (Timm 1996, Sternkopf 1996), reviews (Sternkopf 1996), recipes and knitting patterns (Nordmann 1996), to name just a few. Many of these analyses are even restricted to particular aspects. Only few cover a wider range of genres in a particular subject area. Among them are a few dissertations written at the University of Leipzig (Zerm 1987, Fiedler 1991), Gläser's *Fachtextsorten im Englischen (English LSP Genres)* (1990) as well as my own *Textsorten in Naturwissenschaften und Technik (Genres in Science and Technology)* (Göpferich 1995a) and Munsberg's *Mündliche Fachkommunikation (Spoken LPS Communication)* (1994) in the field of chemistry. The last three are among the most systematic analyses of ranges of genres in written and spoken communication respectively.

Now the time has come to accompany further analyses of individual genres with an integration process – an integration process in which the results of the individual analyses are joined together to form a mosaic that gives us an overview over larger text-type systems. The methods of such an integration will be the subject of the next section.

Furthermore, we should juxtapose current usage as it manifests itself in the genres analyzed with an empirically-based description of the way ideal texts should be structured and formulated. This description, which must be independent of current usage to a certain degree, must define what a text of a particular genre and thus a particular communicative function should look like to be of utmost use for the specified purposes. In the age of globalization, one such purpose may be optimal suitability for machine translation. Other purposes may be rapid accessibility and cognitive processability in a particular communicative situation. Here we have to keep in mind that every-day life and our jobs confront us with new questions every day which we can only answer within a reasonable span of time if we have rapid access to the knowledge required. We can only extract the knowledge we need from the flood of information which inundates us and process it quickly enough if it is structured and presented adequately. The methodology to determine what such ideal genres should look like will be the subject of the second part of this article.

Descriptive LSP text(-type) linguistics: the methodology of investigating current usage

Let us first turn to the methods by which the results of the numerous analyses of specific genres carried out so far can be integrated into a text-type system which gives us insight into the relationships between LSP texts.

When we select the investigations whose results are to be integrated in our text-type system and when we actually integrate them into this system, we have to take into consideration the following aspects:

a) Locating the analyses in a common framework

In order to ensure that the results from individual genre analyses are comparable and that they can be joined together to form a consistent picture of LSP communication, Kalverkämper (1992: 68 ff.) calls for the inclusion of definition sections in all genre analyses. These definitions must specify how each analysis fits into the range of investigations in LSP research. For this purpose, he develops a hierarchically structured framework, which – in an ascending order – comprises the following criteria: language system ("Sprachsystem"), meaning the level of linguistic description (from the level of phonemes, morphemes, and lexical items via the level of phrases, sentences, and sequences of sentences up to the level of text parts, texts, and texts-in-situations/functions); variety (language for general purposes, language for specific purposes in subject area 1, in subject area 2, etc.); vertical level ("vertikale Schichtung"), ranging from the language of consumers up to the level of the theoretical sciences; medium (spoken vs. written); intralingual vs. interlingual/intercultural comparison; and time (synchronic vs. diachronic). This framework was used in both my own (Göpferich 1995a: 10) and Munsberg's (1994: 10) analyses.

If individual analyses whose results are to be integrated into a text-type system have not yet been located in such a framework which serves as a *tertium comparationis*, this should be done first. Such metaanalyses are essential so that, subsequently, the results of investigations which resemble each other in a maximum number of criteria (at best in all criteria except one) can be compared with each other. Only in this way is it possible to pinpoint the reasons for differences discovered in contrastive genre analyses.

If a larger number of genre analyses exist which resemble each other in all aspects of Kalverkämper's framework except one (as is the case in my analyses [Göpferich 1995a], where the genres of written communication differ only in the vertical level they belong to), the results of the analyses and metaanalyses can be used to develop hierarchical typologies for the particular subject area (cf. my pragmatic typology of LSP texts in science and technology [Göpferich

1995a, 1995b]). Hierarchical typologies of this kind have the advantage of allowing us to find out in a systematic way which features are genre-specific and which features a genre shares with one or more other genres (cf. Isenberg 1983: 305, and also Kalverkämper 1992). Isenberg (1983: 305) considers this a prerequisite for a comprehensive and theoretically satisfactory description of a genre.

When developing such a typology, a differentiation criterion has to be selected for each typology level. Here we can make use of the results of communication-oriented genre linguists who, in accordance with Gülich's und Raible's (1977: 25) extended model of linguistic communication, agree "that in genre descriptions 'internal' (linguistic) as well as 'external' (communicative, situational) features have to be considered, but that the former are determined by or even a consequence of the latter"[2] (Lux 1981: 35 f.; my translation). This means that we can proceed on the assumption that the linguistic features of a genre depend on its communicative purpose. Consequently, a communicative-pragmatic typology basis should offer the best chance of reducing the variety of genres to a limited number of text categories and subcategories. This is supported by the results of my analyses (Göpferich 1995a).

b) Intralingual comparisons to discover genre-specific features in contrast to general textuality features

If the objective of contrastive genre analyses is to discover only *genre-specific* features, the analyses have to extend to a maximum number of genres which resemble each other in their communicative function and cover similar topics. Only in this way is it possible to find out which features are shared by similar genres or are even general textuality features, i.e., features shared by all texts, and which ones turn out to be genre-specific (cf. Spillner 1981: 242).

c) Analyses of texts by different authors to identify idiosyncrasies

In order to avoid misinterpreting idiosyncrasies of an author as genre-specific features, genre analyses should always extend to a number of texts by different authors.

d) Restriction to particular subject areas

Conventions do not only vary from genre to genre, but can also differ depending on the subject area. In order to ensure that differences in what are considered genre conventions are only genre-related and not (also) subject-area-related, the texts to be analyzed must belong to the same subject area and deal with similar topics. In the fields of technology and engineering this can easily be achieved since, here, a product becomes the subject of various genres during its development, production and use. Thus, it is possible to collect and compare all texts which accompany a particular product – from its invention (patent specification) through the construction of its first prototype (e.g., article in a learned journal) and the beginning of its mass production (e.g., parts lists and order specifications) until it reaches the customer (product information, operating instructions, etc.) and service personnel (e.g., workshop repair manual).

e) Considering functional diversity within texts

Especially when using statistical methods, we have to take into account that texts are not homogeneous, but consist of sections with different communicative functions. Due to these different functions, the individual sections of a text may vary in the linguistic means used in them and their frequency. As a consequence, in contrastive genre analyses each section has to be analyzed separately, and, subsequently, only the results of those sections may be contrasted which share the same communicative function within the texts' macrostructures (cf. Göpferich 1995a: 9 f., 478; as well as Minogue's and Weber's text-comparison method C [1992: 56 f.]).

f) Intralingual comparisons as a prerequisite for interlingual/intercultural comparisons

Reiß and Vermeer (1984: 192) have emphasized that cultures may differ in the ranges of genres which are conventionalized in them. They differentiate general (classes of) genres, which can probably be found in all cultures, (classes of) genres which may be found in more than one, but not in all cultures, and (classes of) genres which are typical of hardly more than one culture (Reiß and Vermeer 1984: 192). The genres in science and engineering can probably be found in at least all industrialized cultures (cf. Göpferich 1995a: 172); nevertheless even in this range of texts, features can be found in which the genres of two cultures differ.

Kußmaul (1995) provides an example of this: in Germany washing machines come with instruction leaflets which not only instruct the reader on how to use the washing machine but also on how to install it, since the wages for manual labor are very high in Germany and therefore small jobs like installing the machine are usually done by the housewife's husband or even the housewife herself. As a consequence, the installation instructions must be very detailed and comprehensible for a lay reader. For the Indonesian market, however, the information which, in Germany, is combined in one instruction leaflet for laypersons must become the subject of two separate instruction leaflets. Those who can afford a washing machine in Indonesia can also afford to have it installed by the dealer since wages are low in Indonesia. Consequently, the installation instructions must be addressed to a different, more experienced audience. Furthermore, the buyer of the washing machine will never use it himself but leave this to domestic servants. Many of these people, however, cannot read, so that it is no use translating the text into their language. What they need are picture sequences which explain how to use the machine. The Indonesian functional equivalent of a German instruction leaflet for a washing machine thus consists of two instruction leaflets, neither of which may assume the same knowledge base as the German version (cf. Kußmaul 1995: 75).

Such differences can only be detected if interlingual or intercultural genre comparisons are preceded by intralingual ones which result in intralingual text-type systems. The intralingual text-type systems of two languages can then be contrasted interlingually – a method well known from terminology research (cf. e.g. Arntz 1992).

g) Parallel texts as material for interlingual/intercultural comparisons of genres

The material to be compared interlingually or interculturally to detect differences in genre conventions must be *parallel texts*. Parallel texts are texts in different languages and/or from different cultures which are not translations of each other, but cover comparable topics and are written for the same purpose. Spillner (1981: 241 f.) differentiates three types of parallel-text analyses: the comparison of text adaptations ("Vergleich von Textadaptationen"), the comparison of texts for equivalent situations ("situationsäquivalenter Textvergleich") and the comparison of genres ("Textsortenkontrastierung").[3]

Like Reiß and Vermeer (1984: 196) I consider only the last two methods parallel-text comparisons in a stricter sense. As Scheithauer (1987: 34) points out, text adaptations are really pragmatic translations ("Übersetzungen unter pragmatischen Gesichtspunkten"). When using translations in interlingual or intercultural comparisons, one always runs the risk of considering features which are the result of interferences to be target-culture conventions (Reiß and Vermeer 1984: 195). Therefore, translations which are to be used for interlingual or intercultural comparisons must first be subjected to a translation-quality assessment which has to check whether the target culture conventions have been complied with. For such a quality assessment, the genre conventions must be known in advance – which quite often is still not the case.

Comparisons of texts for equivalent situations (cf. Spillner 1981: 242) are nothing else than comparisons of genres, since texts of the same genre are characterized by the fact that they are produced for the same communicative situation and for the same communicative purpose.

h) Comparison at all levels of linguistic description

Since genre conventions and thus intercultural differences in them may occur on all levels of linguistic description (cf. Reiß and Vermeer 1984: 184 f.) and since conventions on different levels may be interdependent, profound genre comparisons must also extend to all levels of linguistic description. This can be achieved by following Hoffmann's principle of "cumulative text analysis" (Hoffmann 1983: 63).

From perpetuation to optimization in LSP text(-type) linguistics

In our mother-tongue classes as well as in foreign language and translation classes, we tacitly assume that imitating the genre features of the respective (target) language which we collect on a purely descriptive basis by analyzing authentic texts in the respective language leads to the production of texts which optimally or at least adequately fulfil their communicative functions (cf. the quotation from Hoffmann above). This seems to be based on the belief that the characteristics we find have developed into genre conventions because they contribute to the communicative function of the respective texts in an ideal way (cf. Göpferich 1995a: 153).

To proceed on this assumption seems plausible; whether the characteristics we find, however, actually serve the communicative purposes of the genres in which they occur in an ideal way has – apart from a few exceptions – not yet been investigated. Among the exceptions are studies by instructional psychologists who, for example, have examined the effect of *advance organizers* (Ausubel 1968), word and sentence lengths, summaries, underlining and the like on text retention (cf. the overview by Groeben 1982 and Christmann 1989), and research into the comprehensibility of legal texts (cf. e.g. Gunnarson 1984, Pfeiffer, Strouhal and Wodak 1987). Another example is *usability testing* of instructive texts (cf. Göpferich 1998a).

The fact that incomprehensible instruction leaflets have become a matter of constant complaint and that courses of study in technical communication have been established to overcome obstructions in the flow of information between specialists and non-specialists shows that there must be discrepancies between what we actually find and what we would like to find. Do the deficiencies we find in texts lie outside the range of what is governed by convention? Are they due to violations of conventions? Or are they due to the fact that their authors cling to conventions which are out of date? As Berkenkotter and Huckin (1995: 6) point out: "Genres [...] are always sites of contention between stability and change. They are inherently dynamic, constantly (if gradually) changing over time in response to sociocognitive needs of individual users."

Even if it should still be acceptable to perpetuate current conventions, we have to take a critical look at them to find out whether they really still serve their purpose. This is especially true in the face of rapidly changing communicative conditions due to recent developments in the field of telecommunication and the media. We cannot rely on the fact that the conventions which have served their pupose in the past and present will still do so in future in a new communicative environment. Do the traditional genres perpetuated by descriptive LSP text linguistics still fulfil our wishes? Or, to put it more precisely: is the adaptation of genre conventions keeping pace with developments in our communicative environment? Are the conventions which helped us to solve a communicative problem yesterday still suitable to do so tomorrow, especially in the face of new encoding and transfer methods? Or do we have to establish new norms, new norms which prospective (and dynamic) LSP text(-type) linguistics has to establish on a scientific basis, such as the insights gained by comprehensibility research? – But before I raise any false hopes, I must make clear that I am not going to and I am not able to answer these questions here. What I am going to do is to discuss the methods which may lead to an answer.[4] One thing is certain: simply describing what we actually find in texts of various genres will not answer our questions, especially since many genres are so new that there has not yet been enough time for new conventions to establish themselves. In fact, many texts in new media are nothing more than efforts on the part of their producers to cope with the new medium technically. Numerous technical as well as software problems often deprive the producers of such texts of the time to think about the best way to encode their message.

In our future research we must define requirements to be met by texts of the various genres and, subsequently, analyze whether the texts we are confronted with every day actually meet these requirements. If they cannot fulfil all requirements at the same time, we must establish a hierarchy of requirements.

Examples of requirements to be met by LSP texts

Requirements to be met by genres can be formulated from different points of view. On the whole, we can distinguish between requirements to be met from the point of view of the producer, and requirements to be met from the point of view of the recipient.

Examples of requirements to be met from the point of view of the producer are ease of making updates and reusability of data in other types of documents. Both requirements have led to the establishment of new conventions. Ease of making updates is achieved, for example, by using pagination by chapter or even section instead of continuous pagination throughout the entire document. Pagination by chapter or section has the advantage that insertions in one chapter or section only entail renumbering the pages of the chapter or section where the insertion occurs instead of having to renumber all the pages in the document which follow the insertion.

Reusability of digital data in other documents without the necessity of extensive reformatting is made possible by *SGML (Standard Generalized Markup Language)* and *Document Type Definitions (DTDs)*. SGML is a language for structuring and tagging texts. It is used to assign each part of a text its function instead of its formatting information. In the case of a headline, this function assigned by SGML is simply that it is a headline, not its typographic appearance (e.g., larger and bold fonts). The way in which the individual sections are joined together and their typographic appearance are defined later in the *Document Type Definition (DTD)*. The advantage of SGML is that text (ASCII code) and formatting information are separated and can be used with various computer systems and programs.

The question that arises here is what the ideal DTD for each genre or type of document must look like, and how much room there should be in a DTD for individual preferences, matters of corporate design, etc.

These questions lead us from the requirements to be met by documents from the point of view of the producers to the requirements to be met by them from the point of view of the recipients. 'Recipients' may be both human beings and machines (e.g., machine translation systems). Let us begin with the latter.

Increasing political and social interdependence of nations and economic globalization have led to a continuous increase in the amount of translations required, which in turn makes machine translation a necessity. The quality of machine translation can be improved by producing source texts which use only those lexical items and grammatical constructions the machine translation system can cope with and by avoiding structures which the machine cannot disambiguate. Here, we have to take into consideration that what is ambiguous for a machine translation system may also cause comprehension problems for a human reader.

A first step towards the solution of these problems are controlled languages. In contrast to artificial languages such as Esperanto, controlled languages are subsystems of natural languages which use only a limited number of lexical items and a well-defined system of grammatical rules. Today, many multinational companies (e.g., Rank Xerox, SAP, Eastman-Kodak, IBM, Digital, ITT, Ericsson) as well as the American Ministry of Defense use types of more or less controlled languages which are tailored to the specific requirements of the individual firms or institutions, especially to the machine translation systems they use (cf. Lehrndorfer 1996: 42 f.).

The best-known controlled language is probably the *Simplified English* used by the *AECMA (Association Européenne des Constructeurs de Matériel Aérospatial)*. It has been used internationally for technical documentation in the aerospace industry since 1986 (Lehrndorfer 1995: 123; cf. Lehrndorfer 1996: 41).

Controlled languages such as the AECMA Simplified English, which have been developed as languages for specific purposes (LSPs) only (they are controlled LSPs or, to be more precise, controlled LSPs in the field of science and technology), can be regarded as a first step towards optimization in LSP text(-type) linguistics. What must be criticized about this step, however, is the fact that – apart from restrictions due to machine translatability – many of the restrictions the AECMA Simplified English makes are more or less arbitrary, i.e., not based on the results of investigations into the cognitive processability of texts. Optimal cognitive processability by specified groups of recipients, however, should be at least one of the prerequisites to be met by controlled languages.

Mentioning cognitive processability, we have already turned to the requirements to be met by texts from the point of view of human recipients. Apart from quick or optimal cognitive processability, such requirements are quick accessability of information, comprehensibility, in the case of instructions usability, and other genre-specific demands such as, for example, entertainment in the case of popular science texts.

Furthermore, specific genres are also subject to legal requirements. Among these genres are patent specifications (cf. Göpferich 1995c) and technical documentation, especially operating instructions (cf. Göpferich 1998b: Chap. 11). These legal requirements are examples of prescription in text production. What is prescribed here (e.g., completeness, correctness, simplicity, clearness, comprehensibility),[5] however, is still too general and needs to be specified.

Whenever we ask what texts of a particular communicative function must look like to be of utmost use *to their (human) recipients*, an answer can only be found empirically, e.g., in usability tests and interviews. First approaches of this kind have been made by psycholinguistic and cognitive LSP research, which constitute the current paradigm of LSP research and which have moved a little closer to the recipient on the pragmatic axis. The cognitive turn in LSP research, which became apparent at the beginning of the nineties, in fact, constitutes the prerequisite for optimization in LSP text(-type) linguistics.

Features which in empirical investigations turn out to improve (cognitive) processability may form the basis for genre-specific text production guides, i.e., prospective genre profiles. These prospective genre profiles may supersede the criteria for text evaluation which, up to now, have almost exclusively been based on what we are used to finding in actual texts. A few such features which influence text processing positively have been found by instructional psychologists such as Langer, Schulz von Thun and Tausch (1993) and Groeben (1982). Another contribution to research into text-processing processes has been made by cognitive scientists. Their research results give us an insight into the reasons why certain textual features improve comprehensibility and allow us to specify the results of instructional psychology (cf. the summary by Christmann 1989).

Both instructional psychologists and cognitive scientists, however, have more or less neglected the communicative functions of the texts whose comprehensibility they tried to evaluate, i.e., they have ignored their genres. Furthermore, the specific requirements to be met by LSP texts have not been taken into consideration adequately: what was measured was the texts' *comprehensibility for the general public* – their *general comprehensibility as a sociocultural characteristic*. This is clearly inadequate for the evaluation of specialist communication. These mistakes must be avoided by future genre linguistics, whose objective must be to define genre-specific and readership-specific text production guidelines, i.e., guidelines geared to the specific communicative purposes of texts (cf. section a). In the following, I will discuss a few methodological issues which should be considered when setting up such text production guides.

Prospective LSP text(-type) linguistics: the methodology of investigating what is needed

As stated in the previous section, requirements to be met by texts can be formulated from different points of view: from the point of view of machine translation, from the point of view of text producers, from a legal point of view, or from the point of view of the prospective readership. In this section I am going to confine myself to the perspective of the readership. The question as to which one of several text variants is most suitable for a specific purpose from this perspective can only be answered on an empirical basis, e.g., on the basis of usability testing. Furthermore, we have to take into account that the requirements to be met by specific genres are subject to change, so that the results of empirical investigations need to be verified from time to time.

a) Definition of the communicative purpose

Testing texts for suitability must always be preceded by an exact definition of a text's communicative purpose and intended readership. Communicative purposes may be rapid processability and knowledge extraction by specialists in a particular field, usability of instructions by laymen, optimal retainability even after short reading times, etc. The readership may, for example, comprise specialists in a particular field, laypersons with a certain educational and/or professional background (engineering background vs. linguistic background, etc.).

b) Production of text variants

In the next step, we must find or produce several (at least two) text variants written for the defined communicative purpose. In doing so, we can make use of the results of descriptive LSP text(-type) linguistics, especially the results of interlingual or intercultural analyses. The intercultural differences revealed by these analyses make us aware of the fact that things can be expressed in a way compeletely different from what is common in our own language and help us to take a critical look at our conventions instead of taking them for granted. Imitating those conventions of other cultures which are not common in the culture in which they are imitated, a process termed *interference* in translation science, is one method of producing text variants.[6]

Alternatively, text variants can be produced by first collecting expressions or grammatical constructions which more or less serve the same communicative purpose and then replacing them with each other. Here, too, we can make use of the results obtained by descriptive LSP genre linguistics. In directive speech acts in German operating instructions, for example, we find imperatives *(Schalten Sie das Gerät ein.)*, imperative infinitives *(Gerät einschalten)*, as well as constructions using *müssen (Das Gerät muß eingeschaltet werden.)* or *sein zu* + infinitive *(Das Gerät ist einzuschalten.)*. In order to determine what position in a text requires what kind of construction for the text to serve its purpose best, we have to produce several text versions differing in the constructions used in the individual positions and then present each version to another test group. Tratschitt (1982: 164 f.), for example, found out that German operating instructions are best suited for laymen if a series of instructions is introduced by an imperative, which addresses the reader directly, followed by imperative infinitives, which have the advantage of being the shortest grammatical means of giving instructions in German.

A third method of producing text variants is first testing one version of a text, for example, by having test persons read the text and asking them to pinpoint and comment on the sections they do not understand, and by subsequently producing variants which differ from the original text in exactly the positions indicated by the test persons.

When producing text variants in our multimedia age, we also have to take into account that we have to find out the most suitable medium (picture, video, sound, text) for each type of information. This means that we have to produce document variants which differ in the allocation of different types of information to the various media. A few investigations of this kind have already been carried out. Experiments by Wintermantel, Laux and Fehr (1989) revealed that test persons solved tasks more rapidly (but also made more mistakes) when given pictorial instructions than when given verbal ones. Stone and Glock (1981) found out that test persons having to assemble a toy model were more successful when given combined pictorial and verbal instructions than when given either verbal or pictorial instructions alone. This suggests that in instructions both media should be used in a complementary way.

c) Testing the document variants

To test which document variant is most suitable for the purpose defined (cf. section a), each document version is presented to a different test group. Subsequently, text reception by each group is compared. To measure text reception, questions can be used (e.g., multiple-choice questions, cloze procedures) either immediately after the test persons have read the text or after an interval to be specified. The first method is more suitable when emphasis is placed on comprehension, not retention; the second, when retention is to be measured. Good retention results, however, seem to be an indicator of comprehensibility as well, since – as we know from our own experience – what has been understood can be retained more easily.

If the purpose of the test is to find out whether a text is comprehensible at all (even if comprehension requires some effort on the part of the reader), questions should be asked whose answers cannot be found in the text itself but which require problem solving processes which can only be performed successfully if the text has been fully understood. In tests of this type, the test persons must be allowed to re-read the text while answering the questions. Genres which can be optimized using this test method are non-instructive texts (non-man/technology interaction-oriented texts Göpferich 1995a: 128 f., 1995b: 312 f.), such as research reports, popular science articles, product informations, etc.

A test method particularly suitable for testing instructive texts (man/technology interaction-oriented texts) is usability testing. It allows pinpointing the parts of a text which need optimization – a diagnostic task which cannot be solved by relying on readability formulas, reading times or propositions reproduced in retention tests (cf. Schäflein-Armbruster 1994: 507 ff.). Redish and Schell (1989: 67 ff.) differentiate three types of usability testing, which all have their specific advantages and disadvantages:

1. "User edits" for short sets of instructions, in which the user is told to read the text aloud and to follow the instructions as closely as possible. By noting where the user stumbles, hesitates, or misreads, the researcher can tell where the instructions need to be clarified (Redish and Schell 1989: 68).

2. "Protocol-aided revisions" in which "typical users are asked to perform tasks similar to those that customers will eventually do with the product" and to talk aloud as they perform the task. The thinking-aloud protocols reveal how the test persons interpret the instructions while they are trying to use them. The questions they ask themselves and their misinterpretations help the researcher in pinpointing text sections which need optimization. Since talking to themselves is embarrassing for some people, it can be useful to have two test persons perform a task together and talk to each other about what they are thinking, thus giving them the feeling of having a conversation. In contrast to method 1, protocol-aided revisions can also be used to check whether the test persons can access exactly the information they need in a specific situation.

3. Beta testing (used in the computer industry) where test persons are given a new product and its documentation to try them out and to report any problems they find and/or to respond to questionnaires. This method has the disadvantage that it does not allow a systematic evaluation of the documentation tested (cf. Redish and Schell 1989: 69 f.).

These tests can be accompanied by questions on the part of the researcher or interviews following the test itself (cf. on these "communicative methods" Schäflein-Armbruster 1994: 508 ff, Pfeiffer, Strouhal and Wodak 1987: Chap. 8). Technical equipment which can be used in such tests are devices for recording eye movements during document reception (cf. Winterhoff 1980) as well as programs for registering the keys pressed while using software in protocol files (cf. Bock 1993: Chap. 3).

In all types of testing, the test persons must be representative of the readership the text was written for. Furthermore, in usability testing the tasks to be performed in the test must be similar to the work real users would actually do with the product. The results of such tests form an empirical basis for text production guidelines which lead to texts tailored to the requirements of our information and communication society. These guidelines may have to supersede existing conventions which have grown out of date. Furthermore, they may serve as a means of orientation for technical communicators and translators, as well as for all those who have to produce and evaluate LSP texts.

Finally, let us return to a macroperspective and look at the development of LSP research from a global point of view; then we see:

- *descriptive LSP text(-type) linguistics* of the eighties and the first half of the nineties, which revealed what variables are available to authors producing LSP texts,
- *cognitive LSP text(-type) linguistics* of the nineties – and certainly also the year 2000 –, which has investigated and will investigate how these variables must be used to obtain suitable texts, and finally
- *prospective (optimizing) LSP text(-type) linguistics*, which transforms the results of cognitive LSP text(-type) linguistics into guidelines for text producers and should thus lead to more effective specialized communication.

Notes

[1] I would like to thank my colleague Ingrid Rose-Neiger for proof-reading this article and for her many valuable comments. Any errors, however, are entirely my own.

[2] Cf. Gülich and Raible (1975: 151) and S.J. Schmidt (1978: 55): "[B]earing in mind that the orientation of communicative text theories is towards verbal texts in contexts of communication, any specification of genres has to respect text internal as well as text external aspects and markers (this hypothesis is emphasized by nearly all scholars working in this field)."

[3] R.R.K. Hartmann (1980: 37 ff.) also differentiates three types of parallel texts which more or less correspond to Spillner's categories.

[4] A forum for the discussion of such problems are the PROWITEC Conferences (**Pro**duktion **wi**ssenschaftlicher **Te**xte mit und ohne **C**omputer).

[5] Cf. DIN V 66055 1988: 2; DIN V 8418 1988: 2; VDI 4500 Blatt 1: 3 ff.

[6] Here we have to take into account that we need to get accustomed to new conventions which have been 'imported' from another language or culture.

References

Adams Smith, D.A. 1987. "Variation in field-related genres". *Genre Analysis and ESP. English Language Research (ELR) Journal* 1: 10-32.

Adolphi, K. 1996. "Eine fachliche Textsorte in ihren Bezügen und Abgrenzungen: die Textsortenvariante 'Extended Abstract'". In Kalverkämper and Baumann (1996): 478-500.

Arntz, R. 1992. "Interlinguale Vergleiche von Terminologien und Fachtexten". In Baumann and Kalverkämper (1992): 108-122.

Ausubel, D. 1968. *Educational Psychology: A Cognitive View.* New York etc.: Holt, Rinehart and Winston.

Baumann, K.-D. 1985. "Ein komplexer Untersuchungsansatz für die Differenzierung ausgewählter Fachtextsorten des Englischen" (Summary). In R. Gläser (ed), *Fachsprachliche Textlinguistik – Vorträge der sprachwissenschaftlichen Arbeitstagung an der Sektion Fremdsprachen der Karl-Marx-Universität Leipzig am 11. und 12. Dezember 1984* (Linguistische Studien 133. Reihe A. Arbeitsberichte). Berlin, 142-143.

Baumann, K.-D. and Kalverkämper, H. (eds). 1992. *Kontrastive Fachsprachenforschung* (Forum für Fachsprachen-Forschung 20). Tübingen: Narr.

Baumann, K.-D. and Kalverkämper, H. 1992. "Kontrastive Fachsprachenforschung – ein Begriff, ein Symposium und eine Zukunft. Zur Einführung". In Baumann and Kalverkämper (1992): 9-25.

Baumann, K.-D. and Kalverkämper, H. 1996. "Curriculum vitae – cursus scientiae – progressus linguisticae. Fachtextsorten als Thema: Zur Einführung". In Kalverkämper and Baumann (1996): 13-34.

Berkenkotter, C. and Huckin, T.N. 1995. *Genre Knowledge in Disciplinary Communication: Cognition – Culture – Power.* New Jersey: Lawrence Erlbaum Associates.

Bhatia, V.K. 1996. "Methodological issues in genre analysis". *HERMES Journal of Linguistics* 16: 39-59.

Bock, G. 1993. *Ansätze zur Verbesserung von Technikdokumentation. Eine Analyse von Hilfsmitteln für Technikautoren in der Bundesrepublik Deutschland* (Technical Writing: Beiträge zur Technikdokumentation in Forschung, Ausbildung und Industrie 1). Frankfurt/M. etc.: Lang.

Bolten, J. et al. 1996. "Interkulturalität, Interlingualität und Standardisierung bei der Öffentlichkeitsarbeit in Unternehmen. Gezeigt an amerikanischen, britischen, deutschen, französischen und russischen Geschäftsberichten". In Kalverkämper and Baumann (1996): 389-425.

Christmann, U. 1989. *Modelle der Textverarbeitung: Textbeschreibung als Textverstehen* (Arbeiten zur sozialwissenschaftlichen Psychologie 21). Münster: Aschendorff.

DIN V 8418 1988. "Benutzerinformation. Hinweise für die Erstellung". Berlin: Beuth.

DIN V 66055 1988. "Gebrauchsanweisungen für verbraucherrelevante Produkte". Berlin: Beuth. (= Translation of the ISO/IEC-Guide 37 "Instructions for use of products of consumer interest". May 1983).

Fiedler, S. 1991. *Fachtextlinguistische Untersuchungen zum Kommunikationsbereich der Pädagogik – dargestellt an relevanten Fachtextsorten im Englischen* (Leipziger Fachsprachen-Studien 1). Frankfurt/M., New York, Paris: Peter Lang.

Fluck, H.-R. 1991. *Fachsprachen. Einführung und Bibliographie*. 4th ed. (UTB 483). Tübingen: Francke.

Gläser, R. 1990. *Fachtextsorten im Englischen* (Forum für Fachsprachen-Forschung 13). Tübingen: Narr.

Gnutzmann, K. and Lange, R. 1990. "Kontrastive Textlinguistik und Fachsprachenanalyse". In K. Gnutzmann (ed), *Kontrastive Linguistik* (Forum Angewandte Linguistik 19). Frankfurt/M.: Peter Lang, 85-116.

Göpferich, S. 1995a. *Textsorten in Naturwissenschaften und Technik. Pragmatische Typologie – Kontrastierung – Translation* (Forum für Fachsprachen-Forschung 27). Tübingen: Narr.

Göpferich, S. 1995b. "A pragmatic classification of LSP texts in science and technology". *Target International Journal of Translation Studies* 7(2): 305-326.

Göpferich, S. 1995c. "Von der Terminographie zur Textographie: Computergestützte Verwaltung textsortenspezifischer Textversatzstücke". *Fachsprache/International Journal of LSP* 17(1-2): 17-41.

Göpferich, S. 1998a. "Möglichkeiten der Optimierung von Fachtexten". In L. Hoffmann, H. Kalverkämper and H.E. Wiegand in cooperation with C. Galinski and W. Hüllen (eds). 1997. *Fachsprachen/Languages for Special Purposes: Ein internationales Handbuch zur Fachsprachenforschung und Terminologiewissenschaft/An International Handbook of Special Languages and Terminology Research*. New York, Berlin: de Gruyter. 1003-1014.

Göpferich, S. 1998b. *Interkulturelles Technical Writing: Fachliches adressatengerecht vermitteln. Ein Lehr- und Arbeitsbuch* (Forum für Fachsprachen-Forschung 40). Tübingen: Narr.

Groeben, N. 1982. *Leserpsychologie: Textverständnis – Textverständlichkeit*. Münster: Aschendorff.

Gülich, E. and Raible, W. 1975. "Textsorten-Probleme". In *Linguistische Probleme der Textanalyse* (Schriften des Instituts für deutsche Sprache 35. Jahrbuch 1973). Düsseldorf: Schwann, 144-197.

Gülich, E. and Raible, W. 1977. *Linguistische Textmodelle – Grundlagen und Möglichkeiten* (UTB 130). München: Fink.

Gunnarson, B.-L. 1984. "Functional comprehensibility of legislative texts: Experiments with a Swedish Act of Parliament". *Text* 4: 71-105.

Hartmann, R.R.K. 1980. *Contrastive Textology: Comparative Discourse Analysis in Applied Linguistics* (Studies in Descriptive Linguistics 5). Heidelberg: Groos.

Hoffmann, L. 1983. "Fachtextlinguistik". *Fachsprache/International Journal of LSP* 5(2): 57-68.

Hoffmann, L. 1985. *Kommunikationsmittel Fachsprache. Eine Einführung* (Forum für Fachsprachen-Forschung 1). Tübingen: Narr.

Isenberg, H. 1983. "Grundfragen der Texttypologie". In F. Daneš and D. Viehweger (eds), *Ebenen der Textstruktur* (Linguistische Studien 112. Reihe A. Arbeitsberichte). Berlin (Ost): Akademie der Wissenschaften der DDR, Zentralinstitut für Sprachwissenschaft, 303-342.

Kalverkämper, H. 1983. "Textuelle Fachsprachen-Linguistik als Aufgabe". *LiLi Zeitschrift für Literaturwissenschaft und Linguistik* 13: 124-166.

Kalverkämper, H. 1992. "Hierarchisches Vergleichen als Methode in der Fachsprachen-Forschung". In Baumann and Kalverkämper (1992): 61-77.

Kalverkämper, H. and Baumann, K.-D. (eds). 1996. *Fachliche Textsorten: Komponenten – Relationen – Strategien* (Forum für Fachsprachen-Forschung 25). Tübingen: Narr.

Kretzenbacher, H.L. 1990. *Rekapitulation. Textstrategien der Zusammenfassung von wissenschaftlichen Fachtexten* (Forum für Fachsprachen-Forschung 11). Tübingen: Narr.

Kußmaul, P. 1978. "Kommunikationskonventionen in Textsorten am Beispiel deutscher und englischer geisteswissenschaftlicher Abhandlungen". *Lebende Sprachen* 2: 54-58.

Kußmaul, P. 1990. "Instruktionen in deutschen und englischen Bedienungsanleitungen". In R. Arntz and G. Thome (eds), *Übersetzungswissenschaft. Ergebnisse und Perspektiven. Festschrift für Wolfram Wilss zum 65. Geburtstag*. Tübingen: Narr, 369-379.

Kußmaul, P. 1995. *Training the Translator* (Benjamins Translation Library 10). Amsterdam, Philadelphia: John Benjamins.

Langer, I., Schulz von Thun, F. and Tausch, R. 1993. *Sich verständlich ausdrücken*. 5th ed. München, Basel: Ernst Reinhardt.

Lehrndorfer, A. 1995. "Kontrollierte Sprache und übersetzungsgerechtes Layout". *doku '95: schreiben – illustrieren – übersetzen. 12 Referate am 6. und 7. November 1995, Leonberg*. (Veranstalter: Fachzeitschrift Technische Dokumentation. Adolph-Verlag), 120-130.

Lehrndorfer, A. 1996. *Kontrolliertes Deutsch. Linguistische und sprachpsychologische Leitlinien für eine (maschinell) kontrollierte Sprache in der Technischen Dokumentation* (Tübinger Beiträge zur Linguistik 415). Tübingen: Narr.

Lux, F. 1981. *Text, Situation, Textsorte. Probleme der Textsortenanalyse, dargestellt am Beispiel der britischen Registerlinguistik. Mit einem Ausblick auf eine adäquate Textsortentheorie* (Tübinger Beiträge zur Linguistik 172). Tübingen: Narr.

Minogue, A. and Weber, S. 1992. "Der Textvergleich als Untersuchungsmethode in der Fachsprachenforschung". In Baumann and Kalverkämper (1992): 49-60.

Munsberg, K. 1994. *Mündliche Fachkommunikation. Das Beispiel Chemie* (Forum für Fachsprachen-Forschung 21). Tübingen: Narr.

Nordmann, M. 1996. "Cooking recipes and knitting patterns: Two minilects representing technical writing". In Kalverkämper and Baumann (1996): 554-575.

Ogden, C.K. 1935. *Basic English*. London.

Oldenburg, H. 1992. *Angewandte Fachtextlinguistik. 'Conclusions' und Zusammenfassungen* (Forum für Fachsprachen-Forschung 17). Tübingen: Narr.

Oomen, U. 1972. "Systemtheorie der Texte". *Folia Linguistica* 5(1-2): 12-34.

Pfeiffer, O.E., Strouhal, E. and Wodak, R. 1987. *Recht auf Sprache. Verstehen und Verständlichkeit von Gesetzen* (NÖ Schriften 5 – Wissenschaft). Wien: Verlag Orac.

Pörksen, U. 1974. "Textsorten, Textsortenverschränkungen, Sprachattrappen". *Wirkendes Wort* 24(4): 219-239.

Raible, W. 1972. *Satz und Text. Untersuchungen zu vier romanischen Sprachen* (Beihefte zur Zeitschrift für Romanische Philologie 132). Tübingen: Niemeyer.

Redish, J.C. and Schell, D.A. 1989. "Writing and testing instructions for usability." In B.E. Fearing and W.K. Sparrow (eds), *Technical Writing: Theory and Practice*. New York: MLA, 63-71.

Reiß, K. and Vermeer, H.J. 1984. *Grundlegung einer allgemeinen Translationstheorie* (Linguistische Arbeiten 147). Tübingen: Niemeyer.

Sager, J.C. et al. 1980. *English Special Languages: Principles and practice in science and technology*. Wiesbaden: Brandstetter.

Schaeder, B. 1996. "Wörterbuchartikel als Fachtexte". In Kalverkämper and Baumann (1996): 100-124.

Schäflein-Armbruster, R. 1994. "Dialoganalyse und Verständlichkeit". In G. Fritz and F. Hundsnurscher (eds), *Handbuch der Dialoganalyse*. Tübingen: Niemeyer, 493-517.

Schamlu, M. 1985. *Patentschriften – Patentwesen. Eine argumentationstheoretische Analyse der Textsorte Patentschrift am Beispiel der Patentschriften zu Lehrmitteln* (Studien Deutsch 1). München: Iudicium.

Scheithauer, R. 1987. *Evaluative Elemente in der didaktischen Interpretation von Musik: eine kontrastive Untersuchung anhand der Textsorte Schallplattenkommentar*. Diplomarbeit. Germersheim: Fachbereich Angewandte Sprachwissenschaft der Johannes Gutenberg-Universität Mainz, Institut für Anglistik und Amerikanistik, SS 1987.

Schmidt, S.J. 1978. "Some problems of communicative text theories". In W. Dressler (ed), *Current Trends in Textlinguistics* (Research in Text Theory/Untersuchungen zur Texttheorie 2). Berlin, New York: de Gruyter, 47-60.

Spillner, B. 1981. "Textsorten im Sprachvergleich: Ansätze zu einer Kontrastiven Textologie". In W. Kühlwein, G. Thome and W. Wilss (eds), *Kontrastive Linguistik und Übersetzungswissenschaft: Akten des Internationalen Kolloquiums Trier/Saarbrücken, 25.-30.09.1978*. München: Fink, 239-250.

Spillner, B. 1983. "Zur kontrastiven Analyse von Fachtexten – am Beispiel der Syntax von Wetterberichten". *LiLi Zeitschrift für Literaturwissenschaft und Linguistik* 13(51-52): 110-123.

Sternkopf, J. (1996): "Vorwort und Rezension: Nahe Textsorten für eine ferne Interaktion". In Kalverkämper and Baumann (1996): 468-477.

Stone, D.E. and Glock, M.D. 1981. "How do young adults read directions with and without pictures?" *Journal of Educational Psychology* 73: 419-426.

Timm, C. 1996. "Das Vorwort – eine Textsorte-in-Relation'". In Kalverkämper and Baumann (1996): 458-467.

Tinnefeld, T. 1996. "Die Apposition im französischen Fachtext des Rechts und der Verwaltung – am Beispiel der Textsorte 'Verordnung'". In Kalverkämper and Baumann (1996): 153-174.

Tratschitt, D. 1982. "Über die Anleitungen, Anweisungstexte verständlich abzufassen". In S. Grosse and W. Mentrup (eds), *Anweisungstexte. Forschungsberichte des Instituts für deutsche Sprache Mannheim 54*. Tübingen: Narr, 159-171.

VDI-Richtlinie 4500 Blatt 1 1995. "Technische Dokumentation. Benutzerinformation". Berlin.

Weise, G. 1980. "Textlinguistische Untersuchungen an englischen Zeitschriftenaufsätzen". *Wissenschaftliche Zeitschrift der Friedrich-Schiller-Universität Jena. Gesellschafts- und sprachwissenschaftliche Reihe* 29(6): 675-679.

Winterhoff, P. 1980. "Zum Zusammenhang von Blickbewegungen und sprachlich-kognitiven Prozessen – ein Überblick". *Psychologische Rundschau* 31: 261-275.

Wintermantel, M., Laux, H. and Fehr, U. 1989. *Anweisung zum Handeln: Bilder oder Wörter? Arbeiten aus dem Sonderforschungsbereich 245 "Sprechen und Sprachverstehen im sozialen Kontext"*. (Bericht Nr. 2). Heidelberg, Mannheim.

Zerm, G. 1987. *Textbezogene Untersuchungen zur englischen Fachsprache der Metallurgie (Schwarzmetallurgie)*. Leipzig: Diss.

Index

abstract 77
academic writing 220
accessability of information 237
accountability 33
actant network theory 27
activity 25
Adams Smith 229
Adolphi 229
advertisement 99, 100, 107
advertising function 77
AECMA Simplified English 237
agency 68
alliteration 130
animal nutrition 45
appealing function 79
appetiser 93
argumentation 41, 42, 105
argumentative schema 42
argument structure 43
assessive 103, 108, 110
Atelsek 103
attributed 103, 104, 108
Ausubel 234
author 186
automotive engineering 10

background 108, 109, 110, 111
balanced sentence 133
basic law 153
Baumann 228
Bazerman 178, 183, 187
Bell 97, 101
Bennett 101

Berkenkotter and Huckin 178, 235
beta testing 241
Bex 98, 99
Bhatia 99, 100, 110, 111, 178
bistructured title 84
Bock 242
Bolten 229

Campbell 195
catalyst 15
catalytic converter 15
causal explanation 104
Christmann 234, 238
civil law 153
Clyne 220
cognitive LSP text(-type) linguistics 242
cognitive processability 229, 237
comment 97, 102, 107
commenting 108
commissive 125
common law 153
communicative effectiveness 79
communicative function 122
communicative method 241
communicative purpose 98, 99, 105, 110, 147, 239
component of title 87
comprehensibility 237
concept 36
Connor 178
context 107, 109
contrast 129

contrastive rhetoric 59
controlled language 237
convention 235, 236, 239
Coulthard 99, 100
cross-linguistic equivalence 7
cumulative text analysis 234

data label 47
declaration 123, 125
descriptive 101, 109
descriptive LSP text(-type) linguistics 242
descriptive markup 202
designating function 77, 79
dialogic construction 30
directive 104
directive function 125
directive speech act 239
discoursal strategy 60
discourse analysis 57
discourse community 45, 72, 99, 148
discursive practice 148
dissertation 41
documentation 3, 4
document type definitions (DTDs) 236
Dudley-Evans 43

ease of making updates 236
Edison 31
editorial we 104
Eggins 100, 110, 111
embedding 106
empirical thought 89
Engeström, Engeström and Kärkkäinen 186
epistemological 3
equivalence 6
evaluation 104
experience 89

expository 99, 100, 102, 106, 107, 109, 110
expressive function 9, 123
eye-catcher 90

factual 103
Faigley 185
Fairclough 187
Fiedler 229
fieldwork 67
first impressions 77, 92
Fluck 228
foregrounding 199
Fowler 97
frame 99
Frandsen 111
French 59
function
  advertising function 77
  appealing function 79
  communicative function 122
  designating function 77, 79
  directive function 125
  expressive function 9, 123
  informative function 9
  metatextual function 79
  nominating function 79
  operative function 9
  phatic function 79, 122
  poetic function 126
  representative function 122
  symbolic function 136
  verdictive function 124
functionalist approaches to translation 209

general comprehensibility as a socio-cultural characteristic 238
general-purpose lexicography 5
generic convention 148
generic integrity 148

INDEX

generic knowledge 147
genre 25, 98, 99, 100, 110, 111, 147, 178, 180, 210
genre analysis 57
genre borrowing 107, 110
genre convention 210, 234
genre-embedding 149
genre-mixing 149
genre profile 214
genre prototype 214
genre-specific text production guide 238
geological setting 59
geology 59
Gerzymisch-Arbogast 4
global cohesion 197
Gläser 229
Gnutzmann 229
gossip column 102
Gould and Lewontin 25
Groeben 234, 238
Gunnarson 234
Göpferich 214, 229, 230, 231, 232, 234, 237
Gülich 231

Halliday 98, 193
Hartley 101
Hasan 98, 193
hedge 52
hierarchical typology 230, 231
Hodge 102
Hoey 100
Hoffmann 228, 234
Hoffmann's five-layer model of LSP 13, 14
Holz-Mänttäri 209
hortatory 99, 100, 106, 107, 109, 110
hyponymy 4
Hönig and Kußmaul 209

iconicity 134
ideational aspect 29
illocution 102, 103, 105, 109, 182
illocutionary 111
individuality 26
information content of title 87
informative function 9
inscription 32, 34
institution 186
instruction leaflet 232, 234
interaction 25, 181
interlingual/intercultural comparison 232, 233
interrupted movement 135
intertextual 30
intertextuality 33
interview 238
intralingual 233
intralingual comparison 231, 232
Isenberg 231

Jones 97

Kalverkämper 227, 228, 230, 231
Katz 101
key-word 77
knowledge extraction by specialists 239
Kress and van Leeuwen 185
Kretzenbacher 229
Kußmaul 209, 217, 228, 229, 232, 233
KWIC analysis 7

Lange 229
Langer, Schulz von Thun and Tausch 238
Latour 27
leader 101
leading article 97, 101
lecture 177, 179

legalese 163
legal requirement 237
legislative genre 149
Lehrndorfer 237
length of title 81, 82
letter to the editor 101, 102
level 3
lexeme 4
lexical chain 127
lexical verb 62, 63
lexicology 3
local cohesion 197
logos 29
Longacre 98, 99, 100, 111
Lux 231

Macintyre 97
Mann 98, 99
manual 101
Martin 98, 100, 101, 110, 111, 178
Matthiesen 98, 99
metacommunicative unit 79
metaphor 129
metatext 79
metatextual function 79
minimal argument 41
Minogue 232
monostructured title 84, 88, 90
move 100, 105, 107, 111
move structure 138
multifaceted concept 11
multi-modal text 185
multiword compound 17
Munsberg 229, 230
Myers 186, 187

narrative 99, 100, 101, 108, 109
Neubert 214
news 102, 107, 108
news report 97, 101, 102
news story 109

nominal compounding pattern 16
nominating function 79
Nord 209
Nordmann 229
normative 103
North American approach to genre 28
noun phrase 67

obituary 108
object 30
official voice 139
Oldenburg 229
onomasiological orientation 5
ontology 30
operating instruction 101, 240
operation 35
operative function 9

pagination by chapter 236
parallel structures 130
parallel texts 214, 233
paraphrase 4
part-genre 59
passive 69
perceptual cohesion 197
persuasive 100, 102, 105, 106, 107, 109, 110
petrology 59
Pfeiffer, Strouhal and Wodak 234, 241
phatic function 79, 122
pictorial and verbal instructions 240
Pilegaard 111
plural presidency 137
poetic function 126
politically correct discourse 169
potential equivalent 7
power 187
Powerpoint™ 182
pragmatic factor 170
pragmatics 26

prediction 104
prescription in text production 237
presentational markup 203
principle of conflict 198
principle of intensity 201
principle of proximity 198
principle of reinforcement 198
principle of similarity 201
principle of symmetry and size 202
Prior 186
private intentions 148
problem of interpretation 164
procedural 99, 100
professional communication 4
professional community 147, 167
professional culture 167
professional discourse 167
professional jargon 163
proposal 104
prospective genre profile 238
prospective LSP text(-type) linguistics 238
prospective (optimizing) LSP text(-type) linguistics 242
protocol-aided revision 241
prototype of text type 214
public discourse 149

Raible 228, 231
range of word 82
rapid processability 239
readership 105, 106, 239
recipe 101
recursion 106
Redish and Schell 241
register 101
regularity 26
Reiß 212
Reiß and Vermeer 209, 228, 232, 233, 234
repetition 4, 127

report 97
reporting 108
representational act 29
representative 103
representative function 122
requirements to be met by genres 236
research article 41
"resemanticise" 4
retainability 239
reusability of data 236
rhetoric 26
rhetorical action 147
rhetorical aspect 90
rhetorical cluster 193
rhetorical feature 127
rhetorical situation 147

sales contract 163
Schaeder 229
Schamlu 228
Scheithauer 233
Schriver 193
Schrøder 99
Schäflein-Armbruster 241
scientific and technical writing 26
scientific research article 59
scientific writing 59
script 99
Searle 103
self-representation 31
semantic component 87
semantic correlation 85
semasiological approach 5
SGML (Standard Generalized Markup Language) 236
Simplified English used by the AECMA 237
Skopos theory 209
Snell-Hornby 213
social form 28
social typification 28

source text 209
speech act 103, 104
Spillner 229, 231, 233, 234
stage 100
Standard Generalized Markup Language (SGML) 236
Sternkopf 229
Stone and Glock 240
structure 203
structure of title 83
Stubbs 98
stylistic aspect 90
sub-genre 9
sub-move 61
subtitle 85
suitability for machine translation 229
Swales 99, 110, 178
symbolic function 136
synonym 15
synonymy 4, 6
syntactic incompatibility 78
syntactic structure of title 82, 85

Tannen 99
target text 209
tax form 29
taxonomy 29
technology 185
temporal sequence 108
tense 64
terminological principle 3
terminology 4
text 98, 101
text act 126
text-internal criteria 9
textlinguistic approach to translation 214
text markup 202
text organization 26
text pattern 210

text type 41, 98, 99, 109, 110, 111, 210
text type profile 214
text-type system 229, 230, 233
textual form 28
texture 203
theory of translatorial action 209
Thompson 98, 99
time 186
Timm 229
title 77
  bistructured title 84
  components of title 87
  information content of title 87
  length of title 81, 82
  monostructured title 84, 88, 90
  structure of title 83
  syntactic structure of title 82, 85
Toulmin 43
translation 32
translation-oriented text typology 212
translation quality 209
translation-quality assessment 233
Tratschitt 240
trials of strength 27
Trosborg 111, 187
typology 230, 231

unusual collocation 130
usability 237, 239
usability test 238
usability testing 241
user edits 241
utterance 25

van Dijk 41, 97
verdictive function 124
vertical layering 12
Vestergaard 98, 99, 100, 105
Vienna school 3
voice 68

Watson and Crick 25
Weber 232
Weise 229
White 98
Wilsher 97
Winsor 186, 187
Winterhoff 241
Wintermantel, Laux and Fehr 240

Zerm 229

In the PRAGMATICS AND BEYOND NEW SERIES the following titles have been published thus far or are scheduled for publication:

1. WALTER, Bettyruth: *The Jury Summation as Speech Genre: An Ethnographic Study of What it Means to Those who Use it.* Amsterdam/Philadelphia, 1988.
2. BARTON, Ellen: *Nonsentential Constituents: A Theory of Grammatical Structure and Pragmatic Interpretation.* Amsterdam/Philadelphia, 1990.
3. OLEKSY, Wieslaw (ed.): *Contrastive Pragmatics.* Amsterdam/Philadelphia, 1989.
4. RAFFLER-ENGEL, Walburga von (ed.): *Doctor-Patient Interaction.* Amsterdam/Philadelphia, 1989.
5. THELIN, Nils B. (ed.): *Verbal Aspect in Discourse.* Amsterdam/Philadelphia, 1990.
6. VERSCHUEREN, Jef (ed.): *Selected Papers from the 1987 International Pragmatics Conference. Vol. I: Pragmatics at Issue. Vol. II: Levels of Linguistic Adaptation. Vol. III: The Pragmatics of Intercultural and International Communication* (ed. with Jan Blommaert). Amsterdam/Philadelphia, 1991.
7. LINDENFELD, Jacqueline: *Speech and Sociability at French Urban Market Places.* Amsterdam/Philadelphia, 1990.
8. YOUNG, Lynne: *Language as Behaviour, Language as Code: A Study of Academic English.* Amsterdam/Philadelphia, 1990.
9. LUKE, Kang-Kwong: *Utterance Particles in Cantonese Conversation.* Amsterdam/Philadelphia, 1990.
10. MURRAY, Denise E.: *Conversation for Action. The computer terminal as medium of communication.* Amsterdam/Philadelphia, 1991.
11. LUONG, Hy V.: *Discursive Practices and Linguistic Meanings. The Vietnamese system of person reference.* Amsterdam/Philadelphia, 1990.
12. ABRAHAM, Werner (ed.): *Discourse Particles. Descriptive and theoretical investigations on the logical, syntactic and pragmatic properties of discourse particles in German.* Amsterdam/Philadelphia, 1991.
13. NUYTS, Jan, A. Machtelt BOLKESTEIN and Co VET (eds): *Layers and Levels of Representation in Language Theory: a functional view.* Amsterdam/Philadelphia, 1990.
14. SCHWARTZ, Ursula: *Young Children's Dyadic Pretend Play.* Amsterdam/Philadelphia, 1991.
15. KOMTER, Martha: *Conflict and Cooperation in Job Interviews.* Amsterdam/Philadelphia, 1991.
16. MANN, William C. and Sandra A. THOMPSON (eds): *Discourse Description: Diverse Linguistic Analyses of a Fund-Raising Text.* Amsterdam/Philadelphia, 1992.
17. PIÉRAUT-LE BONNIEC, Gilberte and Marlene DOLITSKY (eds): *Language Bases ... Discourse Bases.* Amsterdam/Philadelphia, 1991.
18. JOHNSTONE, Barbara: *Repetition in Arabic Discourse. Paradigms, syntagms and the ecology of language.* Amsterdam/Philadelphia, 1991.
19. BAKER, Carolyn D. and Allan LUKE (eds): *Towards a Critical Sociology of Reading Pedagogy. Papers of the XII World Congress on Reading.* Amsterdam/Philadelphia, 1991.
20. NUYTS, Jan: *Aspects of a Cognitive-Pragmatic Theory of Language. On cognition, functionalism, and grammar.* Amsterdam/Philadelphia, 1992.

21. SEARLE, John R. et al.: *(On) Searle on Conversation.* Compiled and introduced by Herman Parret and Jef Verschueren. Amsterdam/Philadelphia, 1992.
22. AUER, Peter and Aldo Di LUZIO (eds): *The Contextualization of Language.* Amsterdam/Philadelphia, 1992.
23. FORTESCUE, Michael, Peter HARDER and Lars KRISTOFFERSEN (eds): *Layered Structure and Reference in a Functional Perspective. Papers from the Functional Grammar Conference, Copenhagen, 1990.* Amsterdam/Philadelphia, 1992.
24. MAYNARD, Senko K.: *Discourse Modality: Subjectivity, Emotion and Voice in the Japanese Language.* Amsterdam/Philadelphia, 1993.
25. COUPER-KUHLEN, Elizabeth: *English Speech Rhythm. Form and function in everyday verbal interaction.* Amsterdam/Philadelphia, 1993.
26. STYGALL, Gail: Trial Language. *A study in differential discourse processing.* Amsterdam/Philadelphia, 1994.
27. SUTER, Hans Jürg: *The Wedding Report: A Prototypical Approach to the Study of Traditional Text Types.* Amsterdam/Philadelphia, 1993.
28. VAN DE WALLE, Lieve: *Pragmatics and Classical Sanskrit.* Amsterdam/Philadelphia, 1993.
29. BARSKY, Robert F.: *Constructing a Productive Other: Discourse theory and the convention refugee hearing.* Amsterdam/Philadelphia, 1994.
30. WORTHAM, Stanton E.F.: *Acting Out Participant Examples in the Classroom.* Amsterdam/Philadelphia, 1994.
31. WILDGEN, Wolfgang: *Process, Image and Meaning. A realistic model of the meanings of sentences and narrative texts.* Amsterdam/Philadelphia, 1994.
32. SHIBATANI, Masayoshi and Sandra A. THOMPSON (eds): *Essays in Semantics and Pragmatics.* Amsterdam/Philadelphia, 1995.
33. GOOSSENS, Louis, Paul PAUWELS, Brygida RUDZKA-OSTYN, Anne-Marie SIMON-VANDENBERGEN and Johan VANPARYS: *By Word of Mouth. Metaphor, metonymy and linguistic action in a cognitive perspective.* Amsterdam/Philadelphia, 1995.
34. BARBE, Katharina: Irony in Context. Amsterdam/Philadelphia, 1995.
35. JUCKER, Andreas H. (ed.): *Historical Pragmatics. Pragmatic developments in the history of English.* Amsterdam/Philadelphia, 1995.
36. CHILTON, Paul, Mikhail V. ILYIN and Jacob MEY: *Political Discourse in Transition in Eastern and Western Europe (1989-1991).* Amsterdam/Philadelphia, 1998.
37. CARSTON, Robyn and Seiji UCHIDA (eds): *Relevance Theory. Applications and implications.* Amsterdam/Philadelphia, 1998.
38. FRETHEIM, Thorstein and Jeanette K. GUNDEL (eds): *Reference and Referent Accessibility.* Amsterdam/Philadelphia, 1996.
39. HERRING, Susan (ed.): *Computer-Mediated Communication. Linguistic, social, and cross-cultural perspectives.* Amsterdam/Philadelphia, 1996.
40. DIAMOND, Julie: *Status and Power in Verbal Interaction. A study of discourse in a close-knit social network.* Amsterdam/Philadelphia, 1996.
41. VENTOLA, Eija and Anna MAURANEN, (eds): *Academic Writing. Intercultural and textual issues.* Amsterdam/Philadelphia, 1996.
42. WODAK, Ruth and Helga KOTTHOFF (eds): *Communicating Gender in Context.* Amsterdam/Philadelphia, 1997.

43. JANSSEN, Theo A.J.M. and Wim van der WURFF (eds): *Reported Speech. Forms and functions of the verb.* Amsterdam/Philadelphia, 1996.
44. BARGIELA-CHIAPPINI, Francesca and Sandra J. HARRIS: *Managing Language. The discourse of corporate meetings.* Amsterdam/Philadelphia, 1997.
45. PALTRIDGE, Brian: *Genre, Frames and Writing in Research Settings.* Amsterdam/Philadelphia, 1997.
46. GEORGAKOPOULOU, Alexandra: *Narrative Performances. A study of Modern Greek storytelling.* Amsterdam/Philadelphia, 1997.
47. CHESTERMAN, Andrew: *Contrastive Functional Analysis.* Amsterdam/Philadelphia, 1998.
48. KAMIO, Akio: *Territory of Information.* Amsterdam/Philadelphia, 1997.
49. KURZON, Dennis: *Discourse of Silence.* Amsterdam/Philadelphia, 1998.
50. GRENOBLE, Lenore: *Deixis and Information Packaging in Russian Discourse.* Amsterdam/Philadelphia, 1998.
51. BOULIMA, Jamila: *Negotiated Interaction in Target Language Classroom Discourse.* Amsterdam/Philadelphia, 1999.
52. GILLIS, Steven and Annick DE HOUWER (eds): *The Acquisition of Dutch.* Amsterdam/Philadelphia, 1998.
53. MOSEGAARD HANSEN, Maj-Britt: *The Function of Discourse Particles. A study with special reference to spoken standard French.* Amsterdam/Philadelphia, 1998.
54. HYLAND, Ken: *Hedging in Scientific Research Articles.* Amsterdam/Philadelphia, 1998.
55. ALLWOOD, Jens and Peter Gärdenfors (eds): *Cognitive Semantics. Meaning and cognition.* Amsterdam/Philadelphia, 1999.
56. TANAKA, Hiroko: *Language, Culture and Social Interaction. Turn-taking in Japanese and Anglo-American English.* Amsterdam/Philadelphia, 1999.
57. JUCKER, Andreas H. and Yael ZIV (eds): *Discourse Markers. Descriptions and theory.* Amsterdam/Philadelphia, 1998.
58. ROUCHOTA, Villy and Andreas H. JUCKER (eds): *Current Issues in Relevance Theory.* Amsterdam/Philadelphia, 1998.
59. KAMIO, Akio and Ken-ichi TAKAMI (eds): *Function and Structure. In honor of Susumu Kuno.* 1999.
60. JACOBS, Geert: *Preformulating the News. An analysis of the metapragmatics of press releases.* 1999.
61. MILLS, Margaret H. (ed.): *Slavic Gender Linguistics.* 1999.
62. TZANNE, Angeliki: *Talking at Cross-Purposes. The dynamics of miscommunication.* n.y.p.
63. BUBLITZ, Wolfram, Uta LENK and Eija VENTOLA (eds.): *Coherence in Spoken and Written Discourse. How to create it and how to describe it.Selected papers from the International Workshop on Coherence, Augsburg, 24-27 April 1997.* 1999.
64. SVENNEVIG, Jan: *Getting Acquainted in Conversation. A study of initial interactions.* 1999.
65. COOREN, François: *The Organizing Dimension of Communication.* n.y.p.
66. JUCKER, Andreas H., Gerd FRITZ and Franz LEBSANFT (eds): *Historical Dialogue Analysis.* 1999.

67. TAAVITSAINEN, Irma, Gunnel MELCHERS and Päivi PAHTA (eds.): *Dimensions of Writing in Nonstandard English.* 1999.
68. ARNOVICK, Leslie: *Diachronic Pragmatics. Seven case studies in English illocutionary development.* 1999.
69. NOH, Eun-Ju: *The Semantics and Pragmatics of Metarepresentation in English. A relevance-theoretic account.* n.y.p.
70. SORJONEN, Marja-Leena: *Recipient Activities Particles nii(n) and joo as Responses in Finnish Conversation.* n.y.p.
71. GÓMEZ-GONZÁLEZ, María Ángeles: *The Theme-Topic Interface. Evidence from English.* n.y.p.
72. MARMARIDOU, Sophia S.A.: *Pragmatic Meaning and Cognition.* n.y.p.
73. HESTER, Stephen K. and David FRANCIS (eds.): *Local Education Order. Ethnomethodological studies of knowledge in action.* n.y.p.
74. TROSBORG, Anna (ed.): *Analysing Professional Genres.* 2000.
75. PILKINGTON, Adria: *Poetic Effects. A relevance theory perspective.* n.y.p.
76. MATSUI, Tomoko: *Bridging and Relevance.* n.y.p.
77. VANDERVEKEN, Daniel and Susumu KUBO (eds.): *Essays in Speech Act Theory.* n.y.p.